DATE DUE

New Developments
in
Industrial Polysaccharides

NEW DEVELOPMENTS
IN
INDUSTRIAL POLYSACCHARIDES

Proceedings of the
Conference on New Developments in Industrial
Polysaccharides held at the Stevens Institute of Technology,
Hoboken, New Jersey, May 31–June 1, 1984

Edited by

V. Crescenzi
Stevens Institute of Technology

I.C.M. Dea
Unilever Research, Colworth Laboratory

S. S. Stivala
Stevens Institute of Technology

Gordon and Breach Science Publishers
New York   London   Paris   Montreux   Tokyo

Gordon and Breach Science Publishers

P.O. Box 786
Cooper Station
New York, NY 10276
United States of America

P.O. Box 197
London WC2E 9PX
England

58, rue Lhomond
75005 Paris
France

P.O. Box 161
1820 Montreux 2
Switzerland

14-9 Okubo 3-chome,
Shinjuku-ku,
Tokyo 160
Japan

TP
979.5
P6
C66
1984

Library of Congress Cataloging in Publication Data

Conference on New Developments in Industrial Polysaccharides (1984: Hoboken, N.J.)
New developments in industrial polysaccharides.

Includes index.
1. Polysaccharides—Industrial applications—Con-
gresses. I. Crescenzi, V. (Vittorio), 1942–
II. Dea, I. C. M. (Iain C. M.), 1943–
III. Stivala, Salvatore S., 1923–   . IV. Title.
TP979.5.P6C66 1984   661'.81   85-5592
ISBN 2-88124-032-1

CONTENTS

PREFACE

Starch is perhaps the most widely distributed substance in the vegetable kingdom. It belongs to the group of carbohydrates in which sugars, gums, and cellulose are included. The principal applications of various starches, such as those derived from rice, maize, arrowroot, potatoes, and wheat, are in (a) industry, e.g., sizing in the textile industry and thickening mordants, (b) food, (c) laundry, and (d) pharmaceutics and cosmetics.

Cellulose, the main ingredient of the cell walls of higher plants, comprises at least one-third of all vegetable matter. Cotton and wood are the principal sources of cellulose of industrial interest. Cellulose and its derivatives have been known for more than a century; the first stable derivatives, nitrate and acetate, were prepared in 1883 and 1869, respectively. These were French products, as was cuprammonium rayon, the first regenerated cellulose fiber, which was introduced in 1890. In 1892, the more important viscose rayon (cellulose xanthate) was developed in England. Celluloid, a homogeneous colloidal dispersion of cellulose nitrate and camphor, was developed in the United States in 1870. Enormous development took place after 1920 in the industrial use of cellulose for the production of rayon, cellophane, plastics, lacquers, and adhesives.

While the exploitation of starch and cellulose was the first stage in the industrial utilization of polysaccharides, a wide range of different polysaccharides currently find commercial applications. These include (a) plant gums and mucilages, e.g., gum arabic, pectin, and locust bean gum, (b) polysaccharides of marine origin, e.g., alginate, carrageenan, and chitin, (c) microbial polysaccharides, e.g., xanthan, dextran, and levan, and (d) ionic polysaccharides from mammalian tissues, e.g., heparin, hyaluronic acid, and chondroitin sulfate. Today, countless industrial companies throughout the world are devoted to the extraction, purification, and modification of these polysaccharides for use in a wide range of applications. It is, therefore, not surprising that there has been a proliferation of technical publications dealing with not only product development and industrial processes but also the fundamental elucidation of the chemistry of these complex polysaccharides, covering biosyntheses, physicochemical characterization, degradation, chemical modification, reaction kinetics and mechanisms, structure determination, and structure-function relationships in biological applications. The last mentioned has led to an increased level of basic and applied research pointing towards emerging applications of polysaccharides in the development of human vaccines and drugs.

Because of these factors, *The International Workshop on New Developments in Industrial Polysaccharides,* held at the Stevens Institute of Technology, Hoboken, New Jersey on May 31 and June 1, 1984, was most timely. The workshop was organized to provide a forum for the communication of results and the exchange of ideas among all the participants. The program was divided into four consecutive sessions, each session having only three invited speakers followed by one hour of open discussions, each being led by a guest discussion leader. The format of the program may be found towards the end of this book. The workshop, chaired in its entirety by Dr. R. L. Whistler, was attended by scientists active in polysaccharide research. The round-table discussions following each session were lively, with short communications from many of the participants. This volume contains the written versions of the materials delivered by the invited speakers and the short communications that evolved from the participants during the open discussions.

The workshop was supported, in part, by the Department of Chemistry and Chemical Engineering and by the Polymer Processing Institute at the Stevens Institute of Technology. We wish to express our appreciation to Dr. Francis T. Jones, Head of the Department of Chemistry and Chemical Engineering, and to Dr. Luigi Z. Pollara, President of the Polymer Processing Institute, for their continued support and encouragement. We extend our thanks to Dr. R. L. Whistler, the invited speakers, the round-table discussion leaders, and the participants, all of whom helped to make this workshop an unqualified success.

The workshop was organized while V. C. was visiting professor at the Stevens Institute of Technology and I.C.M.D. was visiting professor at the Polytechnic Institute of New York.

Vitorrio Crescenzi
Iain C. M. Dea
Salvatore S. Stivala

EARLY HISTORY OF STARCH AND CELLULOSE RESEARCH

H. MORAWETZ
Department of Chemistry
Polytechnic Institute of New York
Brooklyn, N.Y.   11201

In 1811 Gay-Lussac and J. Thénard pointed out that
starch and "ligneous matter" had an elemental composition equal to the sum of carbon and water and suggested
the term "carbohydrate" for such substances.[1] At that
time nothing was known about molecular structure, but
Gay-Lussac noted in 1814 that "the composition of acetic
acid does not differ significantly from that of ligneous
matter. Here are two substances composed of carbon,
oxygen and hydrogen in the same proportions which have
quite different properties. This is a new proof that
the atomic arrangement in a compound has the greatest
influence on its character...Sugar and starch lead to
the same conclusion."[2]

The observation that starch is converted in acid
solution to a sugar was made accidentally in 1811 by
Kirchhoff[3] who, for reasons unknown, hoped to convert
starch into rubber. The discovery of the blue starch-
iodine complex, made only three years after the discovery
of iodine[4] was to prove a valuable analytical tool. In
1833 Biot and Persoz[5] found that the degradation of
starch by acid or an enzyme which they called diastase,
leads to a strongly dextrarotary product, which they

named dextrin. This was at first thought to be an intermediate in the conversion of starch to sugar, but
Musculus[6] showed that dextrin and sugar were produced
concurrently. The sugar resulting from the enzymatic
degradation of starch was identified in 1876 by O'Sullivan[7]
as a disaccharide which he named maltose.

By 1879 Brown and Heron[8] wrote that they had a list
of 400 publications dealing with starch. Research in
this area was favored by the availability of three independent analytical methods employing the starch-iodine
complex, the optical activity and the reducing power.
Brown and Heron concluded that starch molecules contained at least ten maltose residues; these split off
one maltose unit at a time with equilibrium being
attained at 80% maltose and 20% dextrin.

In 1899 Brown and Morris[9] obtained a cryoscopic
molecular weight of 6400 for dextrin, which they considered to be one-fifth of the starch molecule. However,
ten years later the influence of the "colloid chemistry
school" had made Brown and Millar[10] write that "the
method of freezing is not applicable to colloidal
products." Instead they lightly oxidized dextrin to a
"dextrinic acid" and found that the carboxyl content
suggested a chain of forty glucose residues in the
dextrin and 200 in the starch. The distinction between
amylose and amylopectin was clarified only at the
beginning of this century by Maquenne.[11]

Strangely enough, the chain character of starch came
again into disrepute as colloid concepts gained the
upper hand. The influential organic chemist Karrer
claimed[12] that starch is in colloidal solution analogous

to colloidal silver solutions-true solutions are obtained
for starch acetate just as with silver nitrate. He
thought that the disappearance of the starch-iodine color,
"a typical colloid phenomenon" when starch is acetylated
was proof of this concept. As late at 1921, Karrer and
Nägeli[13] argued that it made no sense to assume that
plants would synthesize a polysaccharide which would
have to be degraded again to the sugar used in the
metabolism of the plant.

The history of cellulose research starts with Payen's
discovery[14] that Gay-Lussac and Thénard were wrong in
claiming that wood and starch had ths same elemental
composition, but that after extraction with dilute nitric
acid a substance with an elemental composition identical
to that of starch can be isolated. This substance,
characteristic of cell walls, was named cellulose.
Payen believed that starch and cellulose differed
merely in their states of aggregation whild Frémy[15] ad-
vanced the view that they were isomers.

In the early years "cellulose" was used to include
a variety of polysaccharides since techniques to
differentiate between different sugars were limited.
Flechsig[16] found that hydrolysis of cotton yielded only
glucose and Schulze[17] suggested that the designation of
cellulose be restricted to "constituents of cell walls
resistant to dilute acid and alkali, soluble in ammoniacal
copper solution and yielding grape sugar on hydrolysis."
Skraup and König[18] isolated in 1901 cellubiose octa-
acetate from the acetolysis of cellulose but Hess[19]
claimed as late as 1928 that the concept of cellulose
built up from cellobiose residues was erroneous.

When x-ray crystallography was applied to cellulose
fibers, the crystallographic unit cell was found to be
surprisingly small. Although Polanyi[20] had pointed out
that this cell could accommodate either two cyclic
disaccharides or two disaccharide segments of polysaccharide chains, the second alternative was forgotten
for a number of years. Later, Polanyi recalled that
the fact "that the elementary cell contained only four
hexoses appeared scandalous"[21], since it was incon-
sistent with the physical properties of cellulose. Yet,
as in the case of starch, there was a strong tendency
to assume that cellulose was an association colloid,
and the small crystallographic unit cell was rationalized
by assuming that "cohesive forces attain the order of
magnitude of valence forces so that any distinction
between them ceases to be meaningful".[22] Only in 1926
did Sponsler and Dore[23], who were convinced that a
"satisfactory formula for cellulose must account for
its physical properties", show that the x-ray diffraction pattern of cellulose is consistent with polysaccharide chains. Although their structure was in-
compatible with a chain built up from cellobiose units
(they assumed alternating 1,1 and 4,4 linkages between
anhydroglucose residues) they felt that the chemical
evidence for this structure was inconclusive. However,
Meyer and Mark[24] showed in 1928 that the x-ray structure
is, in fact, compatible with chains built up from
cellobiose residues. Their data could not distinguish
between parallel and antiparallel chains, but the
parallel arrangement was adopted since it seemed unlikely
that a living organism could produce the antiparallel

structure. A remarkable deduction of physical
properties from this crystal structure was realized in
the next few years when Mark[25] calculated the tensile
strength of an "ideal cellulose" (which turned out to
be about eight times the value for the best cellulosic
fibers) while Meyer and Lotmar[26] estimated the elastic
modulus from crystallographic and spectroscopic data.

REFERENCES

1. J. GAY-LUSSAC and J. THÉNARD, in Recherches Physico-
    Chimiques, Vol. 2, p. 268-350 (1811).
2. J. GAY-LUSSAC, Ann.chim.phys., 91 149 (1814).
3. C. KIRCHHOFF, J. Chem., 4 111 (1812)
4. J.J. COLIN and H. GAUTIER DE CLAUBRY, Ann. chim.
    phys., 90, 239 (1814).
5. J.B. BIOT and J.F. PERSOZ, Ann. chim.phys., [2], 52
    72; 53, 73 (1833).
6. F. MUSCULUS, Ann. chim.phys., [3], 60, 203 (1860);
    [4], 6, 177 (1865).
7. C. O'Sullivan, J.Chem.Soc., 25, 579 (1872); 29, 478
    (1876).
8. H.T. BROWN and J. HERON, J. Chem.Soc., 35, 596 (1879)
9. H.T. BROWN and G.H. MORRIS, J. Chem. Soc., 55,
    462 (1889).
10. H.T. BROWN and J.H. MILLAR, J. Chem.Soc., 75, 286
    (1889).
11. L. MAQUENNE and E. ROUX, Ann.Chim.Phys., [8] 9, 179
    (1906).
12. P. KARRER, Helv.Chim.Acta, 3 620 (1920).
13. P. KARRER and C. NÄGELI, Helv. Chim. Acta, 4 263
    (1921).
14. A. PAYEN, Compt. Rend., 8, 51 (1839).
15. E. FRÉMY, Compt. Rend., 48, 361 (1859).
16. E. FLECHSIG, Z. physiol. Chem., 7, 523 (1882).
17. E. SCHULZE, Ber., 24, 2277 (1891).
18. Z.H. SKRAUP and J. KÖNIG, Ber., 34, 1115 (1901);
    Monatsh., 22, 1011 (1901).
19. K. HESS, "Chemie der Zellulose," Akad. Verlagsges,
    Leipzig, 1928, p. 494.
20. M. POLANYI, Naturwiss., 9, 288 (1921).

21. M. POLANYI, In "Fifty years of x-ray diffraction",
    P.P. Ewald, ed., A. Vosthoek, 1962.
22. R.O. HERZOG and K. WEISSENBERG, Kolloid-Z., 37, 23
    (1925).
23. O.L. SPONSLER and W.H. DORE, Colloid Symp. Mono-
    graph, Vol. 4, 174 (1926).
24. K.H. MEYER and H. MARK, Ber., B61, 593 (1928).
25. H. MARK, Scientia, 51, 405 (1932).
26. K.H. MEYER and W. LOTMAR, Helv.Chim.Acta, 19,
    68 (1936).

CAPSULAR POLYSACCHARIDES OF GRAM NEGATIVE BACTERIA

Guy G.S. Dutton
Department of Chemistry
University of British Columbia
Vancouver, Canada, V6T 1Y6

Capsular polysaccharides of Gram-negative bacteria are discussed with particular reference to those of the genus Klebsiella. The polysaccharides from this genus are noteworthy for the diversity of their structural patterns, several of which are illustrated. It is now generally accepted that bacterial polysaccharides have highly regular structures and it is appropriate to think in terms of repeating units which, in the case of Klebsiella polysaccharides, may contain three to seven monosaccharide residues. Bacterial viruses, or bacteriophages, contain highly specific endoglycanases capable of depolymerizing capsular polysaccharides into oligosaccharides, corresponding to the repeating units, which will still contain labile groups such as pyruvate acetals or ester groups, if they are present in the original polymer. These capsular polysaccharides, and their related oligosaccharides, are easily prepared and represent useful models for studying the relationship between chemical structure and viscosity and as starting materials for the preparation of vaccines.

1. INTRODUCTION

Research groups studying the structures of non-mammalian polysaccharides are finding a wealth of interesting data in

the bacterial kingdom. In order to appreciate the significance of current results and to put these investigations
into perspective it is useful to review briefly the history
of polysaccharide structural determinations.

The determination of the primary chemical structure of
polysaccharides dates from the second decade of this
century following the introduction of the Haworth method of
methylation. Substances studied initially were those that
were readily available in large quantities, suitable for
the methodology of the day - fractional distillation of
methylated sugars and the preparation of crystalline
derivatives - and included cellulose, starch and glycogen.
Each of these is a neutral polymer of D-glucose and is,
thus, a homoglycan. There then followed a period when many
plant gums, such as acacia, tragacanth and mesquite gums,
were examined and these were found to be acidic hetero-
glycans, where the acidity is due to uronic acids. Another
type of acidic polysaccharide is represented by that from
algae where the anionic character is caused by sulfate
groups.

During the 1950 and 1960's many research groups shifted their focus to other plant polysaccharides, the hemicelluloses, both those from annual plants and from trees.
It is only within the last 15 years or so that systematic
studies on bacterial polysaccharides have been carried out
by many different groups and the results have given a new
dimension to the meaning of "polysaccharide structure".
Why is this so?

Despite the excellent early work on plant polysaccharides the results were inconclusive in an exact sense since
these types of polymers do not have regular structures and,

therefore, the best that can be achieved is to determine an
"average structure". By contrast, it transpires that bacterial polysaccharides have regular and well ordered structures. This means that it is convenient to think of a
bacterial polysaccharide as being composed of an oligosaccharide unit, e.g. A-B-C or A-B-C etc, which then
                                      D
repeats in a regular manner until a polysaccharide of high
molecular weight (e.g. 10^6 daltons) is obtained. The
letters A, B, C .... represent monosaccharide residues,
that may be the same or different, and the whole assembly
is referred to as a repeating unit. In the case of the
capsular polysaccharides of the genus Klebsiella the
repeating units have between three and seven sugar
residues, depending upon the strain involved.

It is this regularity of structure that now gives new
meaning to the expression "polysaccharide structure" and
sets apart bacterial carbohydrate polymers from those
isolated from plants. This regularity makes possible
physical methods of examination and characterization,
especially nuclear magnetic resonance (n.m.r.) spectroscopy, as explained later. It also facilitates studies
aimed at correlating primary chemical structure with physical properties and the preparation of vaccines.

It should be noted that current work on plant polysaccharides suggests that they may have regions of regularity but, even so, this feature is noticeably less well
developed than in bacterial systems.

2. ANTIGENS

The interests of the present audience are predominantly
non-medical but it must be recognized that it is mainly the
medical importance of bacteria in infectious diseases that
has led to the current interest in their chemical composition. It will, therefore, be helpful to define some terms
before continuing. An antigen is defined as any substance
that elicits the formation of antibodies in an organism.
It may well be that a certain part of the antigen molecule
plays a more important role in this stimulation of antibody
response and is, therefore, known as the immunodominant
group.

Three types of antigens are recognized in connection
with bacterial cells. These are H antigens due to
flagellae, which are protein in nature; O antigens which
are lipopolysaccharides; and K or capsular antigens that
are polysaccharides. The letters H, O and K come from the
German, Hauch; ohne Hauch and Kapsel. If the chemical
composition of any one of these antigens is changed then
the two bacteria will no longer be identical. In order to
specify completely any particular strain of bacteria it is
necessary that the nature of each antigen be given, thus,
E. coli O9:K32(A):H19. The numbers have no particular
connotation and are arbitrarily assigned to distinguish one
antigen type from another.

In the case of the genus Klebsiella, which has no
flagellae, the capsule is thick and effectively masks the
lipopolysaccharide, so that typing is principally on the
basis of the capsular, or K, antigen.

The remainder of this article will be mainly concerned with the capsular polysaccharides of the genus Klebsiella but many of the comments are applicable to other genera.

3. CAPSULAR POLYSACCHARIDES - Klebsiella

Gram negative bacteria of the family Enterobacteriaceae include Salmonella, Klebsiella and Escherichia coli. The first named is often implicated in cases of food poisoning and strains are differentiated on the basis of their O antigens. Klebsiella may cause pneumonia like diseases and are often found in urinary tract infections, while E. coli is best known for causing "travellers stomach" and is also an important causative organism of infantile diarrhea.

The microbiologist is able to distinguish approximately 80 strains of Klebsiella bacteria based on a serological screening of the capsular K antigens. It may be further demonstrated that these antigens are polysaccharide in nature, whence it follows that there must be 80 polysaccharides of different structures. Most of these have now been examined as a result of the cooperative efforts of several groups.

3.1 Composition

At the time the structural investigations were commenced we were fortunate to know the qualitative composition of all the strains thanks to the work of Nimmich[1,2]. He showed that the antigens are acidic polysaccharides wherein the acidity is due to uronic acid, usually $\underline{D}$-glucuronic, but in some 10% of the strains $\underline{D}$-galacturonic acid. Other sugar

units present are some, or all, of $\underline{D}$-galactose, $\underline{D}$-glucose
and $\underline{D}$-mannose while $\underline{L}$-rhamnose or $\underline{L}$-fucose is found in a
large proportion of the strains.

Non-carbohydrate substituents may play an important
part in the immunological behavior of the antigens as well
as influencing the physical properties of the polysaccharides. Two such substituents are 1-carboxyethylidene
groups i.e. acetals of pyruvic acid e.g.

$$\begin{array}{c} O \diagdown \;\; \diagup O \\ C \\ CH_3 \diagup \;\; \diagdown CO_2H \end{array}$$ and

acetyl groups. The latter often occur in non-stoichiometric proportions but the acetal, when present, forms part
of every repeating unit – occasionally every second unit,
e.g. Klebsiella K70. These pyruvate acetals, first[3]
observed in an algal polysaccharide, are a common feature
of capsular antigens and have been found in approximately
45% of the Klebsiella polysaccharides. The stability of
the acetal group is markedly dependent on the mode of
linkage, thus a 4,6-$\underline{O}$-(1-carboxyethylidene)-hexose is
significantly more stable to acid hydrolysis than 2,3-$\underline{O}$-(1-
carboxyethylidene)-$\underline{L}$-rhamnose. In the latter case, even
passage through a cation exchange column at room
temperature may provoke extensive hydrolysis.

3.2   Nuclear Magnetic Resonance

The structural regularity of bacterial polysaccharides
makes these polymers most amenable to study by n.m.r.
spectroscopy. In effect, the spectrum corresponds to that
of the oligosaccharide repeating unit. The high viscosity
of aqueous solutions of many of these bacterial polysaccharides may cause some line broadening and it is also diffi-

cult to obtain complete exchange with D_2O. As a consequence, an HOD signal is always present which, at ambient
temperature, occurs in the spectral region associated with
anomeric protons. Viscosity may be reduced and the HOD
signal shifted upfield by recording the ^{1}H-n.m.r. spectrum
at an elevated temperature (ca 90°).

In proton spectra two regions are of particular
interest; at high field, δ 1.5–2.5, where signals from
methyl protons of 6-deoxysugars, pyruvate acetals and
O-acetyl groups occur, and at low field, δ 4.5–5.5, where
anomeric protons resonate. When a polysaccharide contains a
6-deoxysugar unit(s) or a pyruvate acetal the integral for
these methyl protons provides a useful internal standard.
The signal for an acetyl group is unsuitable since such a
substituent is likely to be non-stoichiometric.

Given the normal $^4C_1$ conformation of a hexose or
deoxyhexose it follows that when such a residue is
α-linked the anomeric proton is equatorial (1) and,

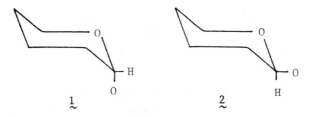

1   2

conversely, when the glycosidic linkage is β, the anomeric
proton is axial (2). Equatorial protons resonate at lower
field than axial and a working rule is that anomeric
signals with δ values above 5.0 correspond to α-linked
units whereas those below δ 5.0 are due to the anomeric
protons of sugar residues with β-glycosidic bonds.

^{13}C-N.m.r. spectra give information similar to that obtained from proton spectra although the distinction between α and β linkages is often less clear. When a poly-saccharide has a disaccharide repeating unit much struc-tural information may be deduced from the chemical shifts of the ring carbon atoms[4,5] but as the complexity of the repeating unit increases this region of the spectrum becomes more difficult to assign. ^{13}C-N.m.r. spectroscopy is particularly useful for detecting the presence of furanosyl units since, in these cases, the signals from anomeric carbons are well downfield (ca 108 p.p.m.) whereas, in proton spectra, signals from anomeric protons of furanosyl and pyranosyl residues overlap[6]. In practice ^{1}H- and ^{13}C-n.m.r. spectroscopy should be considered to give complementary information[7] and, where possible, a polysaccharide sample should be examined by both spectroscopic techniques.

A detailed description of the experimental methods by which chemical structures of polysaccharides are determined is not within the scope of this review. The brief dis-cussion given above of n.m.r. spectroscopy is warranted, since it indicates how readily the anomeric configuration of individual glycosidic linkages may be determined but, more particularly, the quality of the spectra obtained from these bacterial polysaccharide is one of the most compell-ing pieces of evidence for the regularity of their struc-ture and for the concept of a repeating unit.

It is, thus, now appropriate to discuss the structures of some of the capsular polysaccharides that have been obtained from Klebsiella bacteria.

4. STRUCTURAL PATTERNS

It is convenient to consider the structures of the capsular polysaccharides of <u>Klebsiella</u> according to their repeating units and to think in terms of structural patterns. In the diagrams below the symbol X indicates uronic acid (usually $\underline{\underline{D}}$-glucuronic acid) and O any neutral sugar – non-carbo-hydrate substituents are ignored. There are a few capsular polysaccharides lacking any uronic acid where the acidity is due solely to a pyruvic acid acetal – type A. A poly-saccharide may be linear – type B or, if branched, the uronic acid may be a component of the main chain – type C or of the side chain – type D. In each case the number of residues may be from three to seven, depending upon the strain.

Type A
$$- O - O - O - O -$$
$$\begin{array}{c} - O - O - O - O - \\ | \\ O \end{array}$$

K32, K72*
K56

Type B
$$- X - O - O -$$
$$- X - O - O - O - O$$

K1, K5, K63
K70, K81

Type C
$$\begin{array}{c} - X - O - O - \\ | \\ O \end{array}$$
$$\begin{array}{c} - X - O ^- O - \\ | \\ O \end{array}$$

K11, K57
K16, K54

Type D
$$\begin{array}{c} - O - O - \\ | \\ O \\ | \\ X \end{array}$$
$$\begin{array}{c} - O - O - O - \\ | \\ X \\ | \\ O \end{array}$$

K20, K23
K13, K74

* The selection of these and subsequent examples is for the purpose of illustration only and does not claim to be exhaustive.

The two last patterns may be further subdivided; thus, side chains may contain one, two or three residues and, when the uronic acid is a lateral component, it may, or may not, be the terminal residue. Further, in certain instances, one sugar residue may contain two branch points (3)[8] or, alternatively, the repeating unit may contain three single residue side chains (4)[9]. One example has also been reported where the side chain is itself branched (5)[10].

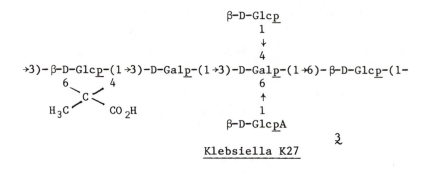

<pre>
                              β-D-Glcp
                                 1
                                 ↓
                                 4
→3)-β-D-Glcp-(1→3)-D-Galp-(1→3)-D-Galp-(1→6)-β-D-Glcp-(1-
        6   4   6
          \ C /   ↑
           >C<   1
       H₃C   CO₂H   β-D-GlcpA
                                                   3
            Klebsiella K27
</pre>

Klebsiella K27

$$→3)-β-D-Glcp-(1→3)-β-D-GlcpA-(1→3)-α-D-Galp-(1→3)-α-D-Manp-(1→$$

<pre>
→3-β-D-Glcp-(1→3)-β-D-GlcpA-(1→3)-α-D-Galp-(1→3)-α-D-Manp-(1→
              4   2   2
              ↑   ↑   ↑
              1   1   1
          α-D-Glcp   β-D-Glcp   β-D-Glcp
                                                 4
              Klebsiella K60
</pre>

<pre>
→3)-α-L-Rha-(1→3)-α-D-Man(1→3)-α-D-Man-(1→3)-β-D-Glc-(1→
                2
                ↑
                1
  β-D-Gal-(1→3)-β-D-GlcA
                4
                ↑
                1
            α-L-Rha
                                   5
          Klebsiella K67
</pre>

A glance at these structures reveals at once the great variation in pattern that one genus of bacteria is capable of producing, which alone makes this area of research a fascinating field of study. Given that any one of the above polysaccharides may be obtained readily in the laboratory in gram quantities, well defined polymers are now conveniently available for studies of the relationships between chemical structure and physical or biological properties (solubility, viscosity, conformation, X-ray pattern, immunological cross reactions etc.).

5. CAPSULAR POLYSACCHARIDES - E. coli

Bacteria of the two genera Klebsiella and E. coli have many properties in common i.e. they are both members of the family Enterobacteriaceae, are of similar morphology and exhibit many of the same biochemical reactions. Despite the importance of E. coli in genetic engineering relatively little is known about the chemistry of its antigens. Some 100K antigens are recognized serologically but even the qualitative composition of most of them is unknown. They have, however, been divided into heat-stable (A) and heat-labile (L) groups[11]. The former type of antigen is considered to be similar in nature to those of Klebsiella. This appears to be true in many cases and, in fact, E. coli K42 has been shown[12] to be identical with Klebsiella K63, and E. coli K30 with Klebsiella K20[13]. Current research, however, reveals that some of the so-called A group antigens contain aminosugars which are noticeably absent from the K polysaccharides of Klebsiella[14].

Some of the L group antigens have been examined and demonstrate the existence of capsular polysaccharides containing sialic acid, fructose, ribitol and other residues not found in <u>Klebsiella</u> K polysaccharides. An earlier review[15] lists the structures of those <u>E. coli</u> capsular polysaccharides known at that time and a more recent compilation is available[16]. Systematic studies on the capsular polysaccharides (K antigens) of <u>E. coli</u> are currently underway both by ourselves and by other groups.

6. <u>SEROLOGICAL CROSS REACTIONS</u>

It has long been known that certain bacteria cross react with heterologous antisera. For example, when rabbit serum containing antibodies against <u>Streptococcus pneumoniae</u> type III is mixed with the polysaccharide from this organism a precipitate is formed as expected, the homologous precipitin reaction. A positive precipitin reaction is also obtained, however, when this serum is mixed with the polysaccharide from <u>Klebsiella</u> K5, this is a cross-reaction or heterologous precipitin reaction. Historically, the structure of PnIII was determined[17] first and was found to be based upon a repeating unit of cellobiouronic acid (6), but there was no chemical explanation for the cross reaction until the structure of <u>Klebsiella</u> K5 was established[18]. This antigen has the structure shown (7)

$$\rightarrow 3)-\beta\text{-D-Glc}\underline{p}\text{A}-(1\rightarrow 4)-\beta\text{-D-Glc}\underline{p}-(-1-$$

6

<u>S. pneumoniae type III</u>

→4)-β-D-GlcpA-(1→4)-β-D-Glcp-(1→3)-β-D-Manp-(1-
               2   4   6
               |
             OAc CH_3 CO_2H

7

Klebsiella type 5

and contains a unit of cellobiouronic acid linked to
D-mannose. The cross-reaction is due to the fact that the
unit of cellobiouronic acid is common to both antigens.
Another example is the cross-reaction between **Klebsiella**
K11 (**8**) and K21 (**9**) where the D-galactopyranosyl residue

→3)-β-D-Glcp(1→3)-β-D-GlcpA-(1→3)-α-D-Galp-(1-
                                 4
                                ↑
                                1

          HO_2C 4
R   ✕   α-D-Galp
         H_3C 6

8

Klebsiella K11

→3)-β-D-Gal-(1→3)-α-D-GlcpA-(1→3)-α-D-Manp-(1→2)-α-D-Manp-(1-
                                4
             ↑   ↑
                                1

        HO_2C 4
R   ✕   α-D-Galp
      H_3C 6

9

Klebsiella K21

carrying the pyruvate acetal is the common feature. It may
be noted that if the pyruvate group is removed both the
homologous and the heterologous reactions fail, indicating
that the pyruvate acetal is the immunodominant group in
these cases.

The structural studies of bacterial antigens in the
last decade have provided a clear, chemical explanation for
many serological cross-reactions previously known only on
an empirical basis. Looked at from the opposite point of
view, if an antibody to a polysaccharide of known structure
cross-reacts with one of unknown structure it is possible
to make certain predictions concerning features likely to
be found in the unknown. Thus, the more structures that
are established unequivocally the greater is the pool of
reference compounds against which antigens of unknown
structure may be screened.

Heidelberger and his coworkers have been the pioneers
in this approach and have made several predictions, subsequently verified chemically, of structural features expected to be found in antigens of hitherto unknown structure.
It should be stressed that the use of serological cross-
reactions as an aid in structural determinations is not
restricted to use within one bacterial genus or even
between bacteria. Thus, Heidelberger and coworkers have
examined cross-reactions within the genus Klebsiella[19];
between Klebsiella, Pneumococci, and Rhizobia[20]; and
between bacteria and plant gums[21]. The generality of serological cross reactions has important implications in the
preparation of vaccines as discussed later.

7. BACTERIAL VIRUSES OR BACTERIOPHAGES

In theory there exists a bacterial virus for every
bacterial system; in practice very many of these are
known. For example, Rieger-Hug[22] and Stirm have been able
to isolate and characterize viruses for about 75% of the

known Klebsiella strains. A few viruses or (bacterio)-
phages have been found for E. coli and one may expect that
many more will be discovered once a systematic search for
them is undertaken. Bacteriophage are often designated by
the symbol ϕ followed by a number corresponding to the
serotype of the host strain; thus, $\phi21$ is the phage for
which Klebsiella K21 is the host strain.

In the present context the most interesting aspect of
those phages that attack the capsular polysaccharides of
Klebsiella is that each contains a highly specific endogly-
canase. The enzyme associated with any one phage will
depolymerize the capsular polysaccharide of the host
strain, but rarely that of another strain and then only
less efficiently[22].

Thus, the enzyme associated with $\phi21$ is a β-galactosi-
dase and will cleave the Klebsiella K21 polysaccharide into
the pentasaccharide corresponding to the repeating unit[23].
Cleavage occurs at the position of the arrow shown in 9.
High yields of complex oligosaccharides, which otherwise
would be tedious to synthesize, may be obtained in this
way. The situation existing with the K21 polysaccharide,
where the pyruvate group is relatively stable to
hydrolysis, may be compared with that of the K32 polymer
(10) where the acetal group is extremely labile. It has

$\rightarrow$2) α-L-Rha$\underline{p}$-(1$\rightarrow$3)-β-L-Rha$\underline{p}$-(1$\rightarrow$4)-α-L-Rha$\underline{p}$-(1$\rightarrow$3)-α-D-Gal$\underline{p}$-(1-

$$3 \diagdown \diagup 4$$
$$\mathrm{H_3C} \qquad \mathrm{CO_2H}$$

Klebsiella K32   10

been demonstrated earlier that the pyruvate acetal is often
the immunodominant group. If it is desired to obtain the
intact oligosaccharide from K32, containing all the
immunological information, e.g. for coupling to a protein
carrier in the preparation of a synthetic vaccine, there is
no way that this could be achieved by acid hydrolysis, even
under mild conditions, since the acetal group would be
cleavage preferentially[23].

The kinetics of these phage mediated depolymerizations
are spectacular. The high viscosity of an aqueous solution
of a capsular polysaccharide is reduced to that of the
buffer within minutes of adding the phage. This behavior
is to be contrasted with β-galactosidases that are commercially available and are quite inactive. Likewise the
enzyme associated with ϕ21 will not depolymerize other
Klebsiella capsular polysaccharides even though they may
contain β-$\underline{\underline{D}}$-galactopyranosyl residues (or will only do so
very slowly). The structural basis for this recognition
phenomenon is, at present, unknown.

The products of such a phage mediated depolymerization
are the oligosaccharide corresponding to a single repeating
unit, sometimes referred to as P1, multiples of these
units, P2, P3 .... etc. and some high-molecular weight
material. The proportions of P1, P2 .... etc. depend to a
large extent on the time for which the depolymerization is
allowed to proceed. Under favorable circumstances yields
in excess of 80% of P1 may be obtained[24]. Oligosaccharides
containing fewer monosaccharide residues than are present
in one repeating unit are not observed. This and the high
yields of P1 obtainable offer further convincing evidence

for the regularity of the structure of these bacterial
polysaccharides.

If one accepts the concept of a repeating unit, the
isolation, following phage depolymerization, of an oligosaccharide representing this unit offers an alternative
approach to structural determinations. Recalling that such
an oligosaccharide (P1) will still contain labile non-
carbohydrate substituents there are certain experimental
advantages to investigating the structure of the oligosaccharide rather than the parent polysaccharide. Chief
amongst these is the parameter of solubility. The oligosaccharide is normally freely soluble in water, thus
affording a non-viscous solution of high concentration
which gives a markedly better signal to noise ratio in
n.m.r. spectroscopy. Similarly, the ready solubility in
solvents such as dimethyl sulfoxide facilitates methylation
analysis. The only information lost by depolymerization of
a polysaccharide to a P1 fragment is the anomeric configuration, in the polymer, of the bond cleaved. This may be
ascertained readily by examination of the [1]H-n.m.r. spectrum of the dimer P2.

8. CONCLUSIONS

The bacterial kingdom produces polysaccharides of diverse
patterns. Many of these materials have high viscosities
and, thus, have potential industrial uses due to their bulk
properties. Some of these aspects have been reviewed elsewhere[25] by Sandford and Baird.

Commercial production of vaccines requires a knowledge
of the primary chemical structure of the antigen(s)

involved and an awareness of any serological cross-reactions. The existence of a cross-reaction can have important practical significance. For example, it may be desireable to prepare a vaccine against bacteria X using the antigen of X. If, however, bacteria X are difficult (or hazardous) to grow the same result may be achieved by using a more convenient system e.g. bacteria Y, providing that there is a serological cross-reaction between the bacteria X and Y.

In any discussion of vaccines it must be appreciated that, in addition to those for human use, there is an important market for farm animals, of which dairy cattle represent but one example.

Due to space limitations only a few structures have been presented in this review but a summary of all known bacterial polysaccharide structures is available[26]. The study of bacterial carbohydrate polymers is an active area of current research which continues to provide exciting results.

9. ACKNOWLEDGMENTS

The work on Klebsiella capsular polysaccharides has involved the international cooperation of several groups. Our contribution represents the combined efforts of many graduate students and research associates to whom I express my gratitude. I am grateful to the Natural Sciences and Engineering Research Council of Canada for continued financial support and to the Killam Memorial Fund for the award of a Senior Fellowship.

REFERENCES

1. W. Nimmich, Z. Med. Mikrobiol. Immunol., 154, 117-131 (1968).
2. W. Nimmich, Z. Allg. Mikrobiol., 19, 343-347 (1979).
3. S. Hirase, Bull. Chem. Soc. Japan, 30, 75-81 (1957).
4. F. -P. Tsui, R. A. Boykins, and W. Egan, Carbohydr. Res., 102, 263-271 (1982).
5. A. K. Bhattacharjee, H. J. Jennings and C. P. Kenny, Biochemistry, 17, 645-651 (1978).
6. G. G. S. Dutton and A. V. Savage, Carbohydr. Res., 83, 351-362 (1980).
7. G. G. S. Dutton and T. E. Folkman, Carbohydr. Res., 80, 147-161 (1980).
8. S. C. Churms, E. H. Merrifield and A. M. Stephen, Carbohydr. Res., 81, 49-58 (1980).
9. G. G. S. Dutton and J. Di Fabio, Carbohydr. Res., 87, 129-139 (1980).
10. G. G. S. Dutton and D. N. Karunaratne, Carbohydr. Res., 119, 157-169 (1983).
11. F. Kauffmann, The Bacteriology of Enterobacteriaceae, (Munksgaard, Copenhagen, 1966).
12. H. Niemann, A. K. Chakraborty, H. Friebolin, and S. Stirm, J. Bacteriol., 133, 390-391 (1978).
13. A. K. Chakraborty, H. Friebolin, and S. Stirm, J. Bacteriol., 141, 971-972 (1980).
14. G. G. S. Dutton, unpublished results.
15. I. Ørskov, F. Ørskov, B. Jann and K. Jann, Bacteriol. Rev., 41, 667-710 (1977).
16. E. Altman, Ph.D. Thesis, University of British Columbia, August 1984.
17. R. E. Reeves and W. F. Goebel, J. Biol. Chem., 139, 511-519 (1941).
18. G. G. S. Dutton and M. -T. Yang, Can. J. Chem., 51, 511-519 (1941).
19. M. Heidelberger, G. G. S. Dutton, J. Eriksen, W. Nimmich and S. Stirm, Acta Path. Microbiol. Scand. Sect. C, 90, 87-90 (1982).
20. M. Heidelberger, W. F. Dudman and W. Nimmich, J. Immunol., 104, 1321-1328 (1970).
21. M. Heidelberger and J. M. Tyler, Immunology, 5, 666-672 (1962).
22. D. Rieger-Hug and S. Stirm, Virology, 113, 363-378 (1981).

23. G. G. S. Dutton, K. L. Mackie, A. V. Savage, D. Rieger-Hug and S. Stirm, Carbohydr. Res., 84, 161-170 (1980).
24. J. L. Di Fabio, G. G. S. Dutton and H. Parolis, Carbohydr. Res., 126, 261-269 (1984).
25. P. A. Sandford and J. Baird, in Polysaccharides Vol. 2, edited by G. O. Aspinall, (Academic Press, N.Y. 1983), Chap. 7, pp. 412-490.
26. L. Kenne and B. Lindberg, in Polysaccharides Vol. 2, edited by G. O. Aspinall (Academic Press, N.Y. 1983), Chap. 5, pp 287-363.

THE EXTRACELLULAR POLYSACCHARIDES OF <u>BACILLUS</u> <u>POLYMYXA</u>

I.C.M. DEA and J.K. MADDEN
Unilever Research,
Colworth Laboratory,
Sharnbrook,
Bedford MK44 1LQ, U.K.

In this chapter we have described the fractionation of
the extracellular polysaccharides of a strain of
<u>B. polymyxa</u> into three chemically distinct polysac-
charides - a minor neutral polysaccharide (N-1), a
minor acidic polysaccharide (A-1) and a major acidic
polysaccharide (A-2).   The separation of the acidic
polysaccharides resulted from a fortuitous stability
of the A-2/CTAB complex.   The neutral polysaccharide
is possibly a cell wall contaminant.   The existence
of two, possibly related, extracellular acidic
polysaccharides from a single strain is unusual,
although not unknown.   The primary structure of the
A-2 polysaccharide has been partially characterised:
it is an acidic polysaccharide containing a mixture of
α- and β-linked residues in the repeating unit;   the
polysaccharide is branched and contains pyruvate as a
substituent.   The results of the rheological studies
have demonstrated that the polysaccharide adopts an
ordered structure in solution, leading to the
formation of a weak gel network.

1. <u>INTRODUCTION</u>

A knowledge of polysaccharide molecular shape and of the
potential polysaccharides have for intermolecular interactions is essential for the understanding and control of
rheological properties.   Polysaccharides can be used to
control the rheology of the aqueous phase in three different ways.   Firstly, they can be used as simple viscosi-
fiers to give shear thinning solutions.   Here the polysaccharide molecules exist as fluctuating disordered chains,
and their viscosity behaviour is non-specific in that when
molecular weight is normalised, a general pattern
describing the concentration dependence and shear dependence of all polysaccharides of this type can be seen.[1]

Secondly, polysaccharides can be used to gel the
aqueous phase.   While thickened polysaccharide solutions
are related to the properties of disordered polymer chains
interacting by virtue of entanglements, the origin of rigid
gel structure is specific, permanent chain-chain polymer
interactions.   Here the relevant interactions are hydrogen
bonding, dipole and ionic interactions, and solvation
terms.   Individually, these interactions are so weak that
conformational stability is achieved only when a large
number of them are simultaneously favourable - that is,
when they act cooperatively to give an ordered
polysaccharide conformation.   For polysaccharides, such
cooperative stabilisation of ordered conformation seldom
occurs for a single chain, but rather requires the
alignment and interaction of two or more covalently regular
chain segments.

Finally, polysaccharides can also control the rheology of the aqueous phase to give properties intermediate between those of thickened solutions and rigid gels.   Such systems have some of the properties of thickened solutions (e.g. reversible shear thinning, pourable, flowable) and some of the properties of gels (e.g. particle suspending).   They are liquids with a finite yield stress.   Usually this occurs because the long polysaccharide chains conformationally order and associate to form rigid rod molecular aggregates, which have the inherent tendency to associate weakly and reversibly to give incipient networks which reform after shear disruption.   These three types of polysaccharide interactions are summarised in Figure 1.

Many bacteria have the ability to produce polysaccharides which lie outside the cell wall.[2] These extracellular polysaccharides may be in the form of a capsule attached to the cell wall or may be secreted as a slime into the growth medium.   There are bacterial extracellular polysaccharides known which exhibit each of the three types of rheological behaviour discussed above.   For example, dextran, the extracellular polysaccharide from Leuconostoc mesenteroides, which is a highly branched 1,6- and 1,3-linked α-D-glucan, confers simple thickening properties to aqueous systems.   In contrast, firm gels are formed by the 1,3-linked β-D-glucan extracellular polysaccharides from Alcaligenes faecalis var myxogenes and certain Agrobacterium species,[3] and the extracellular heteropolysaccharides from Pseudomonas elodes,[4] Arthrobacter stabilis,[5,6]

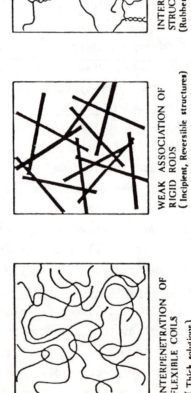
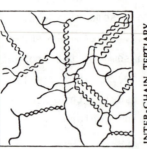
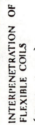

INTERPENETRATION OF FLEXIBLE COILS
(Thick solutions)

WEAK ASSOCIATION OF RIGID RODS
(Incipient, Reversible structures)

INTER-CHAIN TERTIARY STRUCTURES
(Rubbery gels)

FIGURE 1   Different intermolecular interactions in polysaccharide systems

and <u>Arthrobacter</u> <u>viscosus</u>.[5]   In addition, one of the
more actively studied extracellular polysaccharides is
that from <u>Xanthomonas campestris</u>.   This polysaccharide,
xanthan, confers the intermediate, yield stress rheology
to aqueous solutions, and this has led to its widespread
technological exploitation.[7]   The rheological behaviour
of xanthan has been rationalised in terms of the ordered
conformation adopted and the consequent intermolecular
associations.[8]

Our particular interest in the extracellular polysaccharides from <u>Bacillus polymyxa</u> (Strain NCIB 11429)
arose when examination by high-performance gel permeation
chromatography indicated that the material was not
homogeneous but contained two components.[9]   Preliminary
examination of the material by NMR relaxation to obtain
the spin-spin relaxation time, T_2, also provided
evidence for two components.   One component had a T_2
in the millisecond region, typical of a flexible
disordered polysaccharide, and the other had a T_2 in
the microsecond region, indicative of an ordered
structure.   Furthermore the extracellular polysaccharide
preparation, when dissolved in water, exhibits temperature
dependent yield stress solution and gel-forming
properties, depending on the concentration used.   An
understanding of the molecular origins of such interesting
rheological properties requires a knowledge of the primary
chemical structure.   This paper reports preliminary
studies on the extracellular polysaccharides of <u>B. poly-</u>
<u>myxa</u>.   More detailed structural and rheological studies
will be reported elsewhere.[10]

I.C.M. DEA and J.K. MADDEN

2. RESULTS AND DISCUSSION

2.1 Preliminary Characterisation

The extracellular polysaccharide preparation was
hydrolysed and the components examined by paper chromatography of the free sugars and gas liquid chromatography
of the derived alditol acetates. The results indicated
the presence of uronic acid and the neutral sugars fucose,
mannose, galactose and glucose. Amino sugars were not
present. 'H-NMR showed a complex anomeric region with
signals indicative of α- and β-linkages, two doublet
signals arising from the methyl protons of a 6-deoxy sugar
and singlet signals typical of pyruvate and possibly an
O-acetyl substituent.

Extracellular microbial polysaccharides, as a
consequence of their biosynthesis, are composed of simple
regular repeating units[2]. Although these repeating
units are generally small, several larger ones have been
reported including an 11 sugar repeating unit.[11]
However, such large units are rare and consequently should
be treated with caution. Thus, the ratios obtained by
sugar analysis and by a comparison of the anomeric signals
in the NMR spectrum should, in general, be consistent with
this type of simple structure. As the analysis of the
extracellular polysaccharide preparation from B. polymyxa
indicates a very complex repeating unit, this further
suggests that the sample is heterogeneous.

2.2 Fractionation of the Polysaccharide Preparation

Because the analysis showed the presence of both neutral
and acidic sugars, we initially attempted a separation
by ion exchange chromatography.   However, this proved
unsuccessful.   In particular, it was apparent that much
of the material was not being recovered from the column.

The well-known ability of quaternary ammonium salts
to separate mixtures of neutral and acidic polysaccharides[12] was then employed.   These quaternary ammonium
compounds, e.g. cetyltrimethylammonium bromide (CTAB)
and cetylpyridinium chloride (CPC), can form complexes
with acidic polysaccharides causing them to precipitate
while leaving neutral polysaccharides still in solution.
The precipitated complex can generally be disrupted by
use of excess salt, e.g. sodium chloride.   Figure 2 shows
the fractionation scheme using CTAB.   The addition of
a solution of CTAB to a solution of the polysaccharide
preparation caused a precipitate to form immediately.
After centrifugation, the supernatant was concentrated
and polysaccharide precipitated with ethanol, redissolved,
dialysed and freeze dried to give the neutral fraction
N-1.   The precipitate of the acidic polysaccharide/CTAB
complex was treated initially with 2M NaCl at room temperature.   However, the complex remained insoluble.
Complete dissolution could not be obtained even by
stirring in 4M NaCl at 37° overnight.   At this stage the
mixture was centrifuged to give a clear supernatant and
a soft (floppy) gel.   The supernatant was treated with
ethanol, the resulting precipitate isolated, dissolved,
dialysed and freeze dried to give a fraction termed A-1,

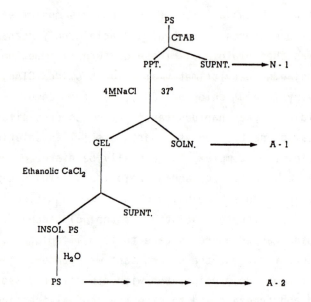

FIGURE 2   Fractionation of B. *polymyxa* extracellular polysaccharides

the first acidic fraction.

As the N-1 and A-1 fractions accounted for only about 25% of the original polysaccharide preparation, it seemed that the remaining gel contained the majority of the poly-saccharide.   It was now necessary to find a method of disrupting the complex to render the polysaccharide soluble.   Scott, in a description of the use of quater-nary ammonium compounds to fractionate polysaccharides, has described several alternative methods to recover the polysaccharide from the complex.[12]   One of these has been applied to a CPC-carrageenate complex which apparently does not dissolve in salt solution.   It involves shaking the complex with an absolute ethanolic solution (saturated) of inorganic salt (e.g. $CaCl_2$ or NaOAc).   Ion exchange takes place with the quaternary ammonium ions entering the alcohol and the Ca (or Na) salt of the polysaccharide remaining insoluble.   This method was applied to the gel.   The Ca salt of the polysacch-aride was dissolved in water and isolated as the second acidic fraction, A-2.

A preliminary characterisation of the individual fractions is shown in Table 1.   It is immediately obvious that three chemically distinct polysaccharides have been obtained.   In particular, the A-1 fraction contains fucose; this is apparent from g.l.c.-m.s. of the alditol acetates and from the 'H-NMR spectrum where two doublet signals are evident at $\delta 1.31$ and $\delta 1.20$ arising from the methyl protons of the 6-deoxy group.   The A-2 fraction, on the other hand, has no fucose but does contain pyruvic acid ($\delta 1.45$ in the 'H nmr spectrum).   The neutral fraction N-1 is a minor component, and by analogy to

TABLE 1   Preliminary characterisation of the polysaccharides obtained by a fractionation of the total extracellular polysaccharide preparation from B. polymyxa

Fraction	Relative Amount (%)	$[\alpha]_D$	Neutral Sugar Composition				Uronic Acid (%)	^{1}H-NMR Chemical Shift (δ)	Assignment
			Fuc	Man	Gal	Glc			
N-1	7	+18°	-	0.8	1	0.3	-	2.08(s)	CH$_3$ of O-Acetyl
A-1	18	+10°	1.2	0.7	1	1.7	22	1.31(d),1.20(d)	CH$_3$ of Fucose
A-2	75	+67°	-	1.7	1	2.7	13	1.45(s)	CH$_3$ of Pyruvate

(s)   singlet

(d)   doublet

other studies,[13] probably arises as a cell wall contaminant; it was not examined further in this study.

2.3 Characterisation of the A-2 Fraction

As the A-2 fraction was by far the major polysaccharide
and as only it formed very viscous solutions (N-1 not at
all viscous, A-1 only slightly so), further
characterisation was concentrated on this polysaccharide.
In order to identify the uronic acid present, the polysaccharide was carboxyl reduced,[14] and the results are
shown in Table 2.   The neutral sugar composition was
examined before and after reduction.   It is evident that
there has been a significant increase in the amount of
glucose, suggesting one glucuronic acid residue per
repeating unit.   This is consistent with the results of
the carbazole assay.   There is some increase in the level
of mannose.   That glucuronic acid was the only uronic
acid present was confirmed by g.l.c. of the TMSi derivatives of the aldonolactones.   The increase in the level
of mannose detected in the carboxyl-reduced sample might
imply that the uronic acid is glycosidically linked to
mannose.   It is well known that the glycosidic linkage
of uronic acids is difficult to hydrolyse.   Thus glucuronic acid linked to mannose might not release all the
mannose on hydrolysis.   However, hydrolysis following
reduction of glucuronic acid to glucose should result in
the release of all the mannose.

Further structural details were elucidated by high
resolution NMR spectroscopy.   A preliminary examination

TABLE 2   Comparison of the neutral sugar composition of native and carboxyl-reduced A-2.

Polysaccharide	Mannose	Galactose	Glucose
		Molar Ratio	
Native	1.7	1	2.7
Carboxyl-Reduced	1.9	1	3.7

TABLE 3   High resolution 'H-NMR data for depolymerised A-2

Chemical Shift (δ)	Coupling Constant J1,2 (Hz)	Integral proton	Assignment
5.37	~3	1	α-anomeric
5.18	~3	2	α-anomeric
4.83	7	2	β-anomeric
4.70	8	1	β-anomeric
4.53	7	1	β-anomeric

indicated the presence of both α- and β- linkages in the polysaccharide.   However, the anomeric signals were very broad and, therefore, in order to improve the resolution the polysaccharide was depolymerised by hydrolysis with acid.   The object was to depolymerise the polysaccharide without causing significant degradation.   The use of 0.1 <u>M</u> trifluoroacetic acid at 95°C for 0.5 h appeared optimum.   Anomeric signals integrating for at least seven different protons could then be distinguished in the spectrum (Table 3).   Two signals, integrating for three protons, at δ 5.37 (1H) and δ 5.18 (2H) were assigned to α-linked residues on the basis of their chemical shifts and small coupling constants.   At least four residues were judged to be β-linked due to their larger coupling constants ($J_{1,2}$ 7-8 Hz) and to their chemical shifts at δ 4.83 (2H), 4.70 (1H) and 4.53 (1H).   Pyruvic acid, which was identified by 'H-NMR, was quantified using a colorimetric assay[15] and found to be present at a level of 3.7%.

Complete methylation of the A-2 polysaccharide has so far proved difficult to achieve.   Methylation analysis by g.l.c.-m.s. of the partially methylated alditol acetates[16,17] derived from both the native and carboxyl-reduced polysaccharide indicates that the polysaccharide is branched with mannose and glucose as terminal sugars, 3,4-linked glucuronic acid and mannose as branch points and considerable amounts of 3-linked glucose. Further analysis should allow a more comprehensive description of the structural details of the repeating

unit including the mode of linkage of the galactose
residue and the location of the pyruvate substituent.

2.4 Heterogeneity of the A-1 Fraction and the Relationship Between the A-1 and A-2 Fractions

Whilst the neutral sugar ratios of the A-2 fraction
remained virtually constant for several different samples
of the polysaccharide preparation analysed, there was a
significant variation in the sugar ratios of the A-1
fraction (Table 4) which suggested that the A-1 polysaccharide was not homogeneous. Although we had already
observed that attempts to chromatograph the biopolymer
on an anion-exchange column resulted in an irreversible
binding of the majority of the polysaccharide to the
column, we still felt that this technique should help to
provide an insight into this problem. Accordingly, a
sample of the A-1 fraction was chromatographed on DEAE-
cellulose. Acidic polysaccharide was eluted with a
continuous gradient of 0-0.5 $\underline{M}$ NaCl, and emerged as two
distinct peaks when monitored by optical rotation. The
first peak, Peak 1, was eluted as a narrow peak at a salt
concentration of 160 mM. This was immediately followed
by another sharp peak, Peak 2, which began to elute at
about 170 mM salt. The analysis of these sub-fractions
is shown in Table 5. A problem with this chromatographic
separation was that only 40% of the material applied to
the column was eluted.

The severe loss of material on chromatography of the

TABLE 4   Variation in the neutral sugar composition of A-1

Sample	Fucose	Mannose	Galactose	Glucose
1	1.0	1.0	1.0	1.0
2	0.8	5.2	1.0	7.3
3	1.2	0.9	1.0	2.0

TABLE 5   Analysis of the polysaccharides isolated as the A-1 fraction and purified by subsequent chromatography on DEAE-cellulose

Frac-tion	Recovery (%)	$[\alpha]_D$	Molar Ratio (Mole %)			
			Fuc	Man	Gal	Glc
Peak 1	20	-69°	1.3(39)	0.1(2)	1(26)	1.3(33)
Peak 2	13	+22°	1.8(25)	1.9(23)	1(12)	3.3(40)

A-1 fraction on DEAE-cellulose makes a precise
interpretation of the results more difficult.   Nevertheless, the results do suggest a connection between A-1 and
A-2.   This is more apparent when pertinent signals in
the 'H-NMR spectra are compared (Table 6).   All of the
polysaccharides showed a complex series of doublet
signals, arising from β-linked sugar residues, resonating
from δ 4.49-4.86.   However, due to inadequate resolution
of the spectra, it would not be legitimate to include
these in the comparison.   Therefore only the singlet
anomeric signals and signals arising from the CH_3 of
the pyruvate and the CH_3 of fucose will be considered.

The spectrum of A-2 is characterised in the anomeric
region by sharp singlets at δ 5.37 and 5.18.   The A-1
spectrum is more complex and contains sharp singlets at
δ 5.67, 5.37 and 4.89 and a broad signal at δ 5.22.   A-1
can be separated by anion-exchange chromatography into
two polysaccharides, isolated as Peak 1 and Peak 2.   Peak
1 has sharp singlets at δ 5.67, 5.22 and 4.89, none of
which appear in the spectrum of A-2.   Peak 2, however,
contains intense signals at δ 5.38 and 5.20 (broad), both
seen in the spectrum of A-2, and weak signals at δ 5.68
and 4.90.   The resonances due to the methyl groups
associated with the polysaccharides can also be compared.
The spectrum of A-2 has a strong pyruvate (CH_3) signal
at δ 1.45.   This signal is present only as a trace in the
spectra of A-1, Peak 1 and Peak 2.   Both the A-1 and
Peak 1 spectra contain strong doublets at δ 1.31 and 1.20
associated with the CH_3 of fucose.   The occurrence of

TABLE 6   Comparison of pertinent signals in the 'H-NMR
spectra of the acidic polysaccharides of
B. polymyxa

A-2	A-1	Peak-1	Peak-2
-	5.67	5.67	5.68(w)
5.37	5.37	-	5.38
5.18	5.22(b)	5.22	5.20(b)
-	4.89	4.89	4.90(w)
1.45	1.45(tr)	1.45(tr)	1.45(tr)
-	1.31(d)	1.31(d)	1.31(d),(w)
	1.20(d)	1.20(d)	1.20(d),(w)

(w)   weak signal

(b)   broad signal

(tr)   trace

(d)   doublet

two distinct doublet signals implies that fucose is
present in two different chemical environments within the
polysaccharide.   These doublets are much less intense
in the spectrum of Peak 2 and presumably arise due to
incomplete separation from the polysaccharide of Peak 1.

The combined evidence of sugar analysis and NMR
characterisation suggests that A-1 is a mixture of two
acidic polysaccharides, a fucose-containing polysaccharide
devoid of mannose (Peak 1) and non-pyruvylated A-2
(Peak 2).   This latter material still contains small
amounts of the fucose-containing polysaccharide.   It
would, of course, be desirable to obtain independent
chemical evidence, such as that afforded by a methylation
analysis of the individual polysaccharides, to corroborate
this.   The nature of the polysaccharide which remains
bound to the ion-exchange column is not certain.
However, it is unlikely that it could be a third structur-
ally-unrelated polysaccharide as a comparison of the
'H-NMR spectra suggest that Peaks 1 and 2 together account
for all the signals present in the spectrum of A-1.

The separation of A-1 and A-2 was achieved due to
their different stabilities within the polysaccharide/CTAB
complex.   As A-1, apparently containing non-pyruvylated
A-2, can be readily isolated from the complex using excess
salt, it may be that the pyruvate present in A-2 plays
an important role in stabilising the complex.   It may
fulfil this function by increasing the charge density of
the polymer or, conceivably, by influencing the conformation adopted by the polysaccharide, thereby leading to
a 'tighter fit'.   The pyruvate groups of xanthan have

also been shown to promote intermolecular association as
a consequence of apolar interactions between the pyruvate
methyl groups.[18]

2.5 The Origin of the Two Acidic Extracellular
Polysaccharides of B. Polymyxa

The strain of B. polymyxa used in this study clearly
produces two structurally different acidic extracellular
polysaccharides, namely, A-2 and the fucose-containing
polysaccharide of A-1.   The other polysaccharide of A-1
appears to be non-pyruvylated A-2.   Although the production of more than one extracellular polysaccharide by a
pure bacterial strain is not usual, it does seem from
recent literature that it is more common than was once
believed.   Thus a pure strain of Rhizobium meliloti
(strain 201) has been reported to produce two extracellular acidic polysaccharides.[19]   The purity of the
strain was ascertained by repeated single-colony isolation.   Similarly, the extracellular mucilage of
Beijerinckia mobilis has been shown to contain small
levels of neutral component together with a major acidic
polysaccharide.[20]

During continuous culture studies on the production
of extracellular polysaccharides by a strain of Xanthomonas
juglandis Evans et al.[21] noted that the bacteria
produced two extracellular polysaccharides, one similar
in composition to xanthan and the other composed of
glucose and rhamnose.   The ability of mutants of
X. campestris to produce extracellular polysaccharides
differing in composition to the xanthan of the wild-type

has been demonstrated.[22]   One type of mutant (crenated
mutant) produced an extracellular polysaccharide
containing significant amounts of galactose and rhamnose
in addition to the normal components of glucose, mannose
and glucuronic acid.   Samples of extracellular
polysaccharide from the same crenated strain grown under
identical conditions had different compositions, unlike
the wild-type strain which was remarkably constant in its
composition from batch to batch.

Strains of Alcaligenes faecalis var myxogenes can
produce the β 1 - 3 glucan curdlan in addition to a
succinoglycan-type heteropolysaccharide[3].   Similar
behaviour has been found in certain Agrobacterium species
which can undergo a spontaneous mutation to produce
curdlan and succinoglycan.[3,23]   A strain of Rhizobium
trifoli also mutates spontaneously to produce a succino-
glycan-like polysaccharide and curdlan.[24]   Thus, the
ability of bacteria to undergo a spontaneous mutation
resulting in the production of polysaccharides differing
in structure to those of the wild-type strain has been
well documented.

Extracellular polysaccharides of B. polymyxa have
been reported.   Ninomiya and Kizaki[25] have isolated
an acidic heteropolysaccharide composed of glucose,
mannose, galactose and glucuronic acid in the molar ratios
of 3 : 3 : 1 : 2.   An apparently different strain of
B. polymyxa isolated by Jeanes and her colleagues produced
a polysaccharide which was closely similar in chemical
composition.[26]   Pyruvate was present as a substituent.
A strain of B. polymyxa has also been found which produces

a fucose-containing neutral extracellular polysaccharide.[27]

The strain of <u>B. polymyxa</u> used in our study was obtained by repeated single-colony isolation. Purity was assessed on the basis of colonial and microscopic morphology. The strain was noticed to produce two colony types differing in gross colonial appearance during growth on solid media: s type colonies which were smooth, gummy and translucent with few spores and a round, even edge; k type colonies which were crinkly, less gummy and opalescent/white, containing many spores and an undulated edge. Further propagation from a single colony still gave rise to both types. It is not yet clear whether such a variation in colonial morphology is due to a regular spontaneous mutation or to a phenotypic response and, further, whether the occurrence of two extracellular anionic polysaccharides is a manifestation of this variation.

2.6 Rheological Studies

Solutions of the whole polysaccharide preparation are very viscous at very low concentrations. For example, an 0.5% w/v solution has a viscosity of 86 poise at a shear rate of 1 sec^{-1} which compares with a viscosity of 40 poise for 0.5% w/v xanthan at this same shear rate. At concentrations above $\sim 0.5\%$ w/v, the total polysaccharide preparation forms thermally-reversible gels. We have carried out a series of rheological studies on the polysaccharide using techniques which in the past have

proved useful in the characterisation of intermolecular
associations.   In the main these studies have been on
the whole polysaccharide preparation rather than on the
A-2 fraction.   However, the indications are that,
although it is the A-2 fraction which is the active
ingredient, the rheological behaviour of the total mixture
is very similar, if not identical.

The response of solutions of disordered and ordered
polysaccharides differs when a stress is applied.
Solutions of disordered, random coil polysaccharides
continue to flow as the stress increases.   Solutions of
ordered polysaccharides such as xanthan require a finite
stress before flow can begin.   The point at which this
occurs is known as the yield stress.   This behaviour has
been interpreted in terms of an intermolecular association
of rigid rod molecules.[8]   Below the yield point the
ordered polysaccharides are considered to be associated
by noncovalent interactions to form a weak gel network.
As the yield point is reached this network is broken and
the ordered molecules begin to align themselves in the
direction of flow resulting in a rapid decrease in
viscosity (shear thinning behaviour).

A consequence of the existence of a yield stress in
the solution is seen if oscillatory measurements are made
and dynamic viscosity is plotted against frequency.
Thus for xanthan (Figure 3), once the system begins to
move there is a rapid decrease in dynamic viscosity with
increasing frequency as the rigid molecules align
themselves in the direction of flow.[8]   This decrease is

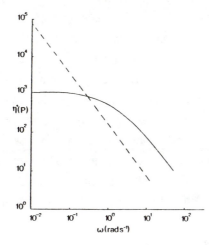

FIGURE 3   Typical flow curves of an ordered
            polysaccharide (---) (e.g. 2% w/v xanthan)
            and a disordered polysaccharide (——) (e.g.
            5% w/v λ -carrageenan)

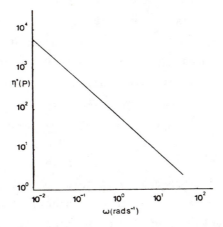

FIGURE 4   Plot of dynamic viscosity (η^*) versus
            frequency (ω) for a 1% w/v total
            polysaccharide preparation from <u>B. polymyxa.</u>

much greater than for random coil polysaccharides in
solution, as shown for λ-carrageenan in Figure 3.
There is a flat region of constant viscosity (the
Newtonian region) for random coil polymers which occurs
at low frequency.   In this region intermolecular
entanglements which are disrupted by the imposed deform-
ation are replaced by new interactions between different
partners.   As there is no net change in the extent of
entanglement, there is no change in solution properties.
However, when the rate of externally imposed movement
becomes greater than the rate of formation of new
entanglements a rapid decrease in dynamic viscosity
occurs.   The experimental data for the Na salt of the
total polysaccharide preparation (1% w/v in water) is
shown in Figure 4.   As can be seen, the curve has the
form of an ordered polysaccharide in solution with no
evidence of a region of constant viscosity.   With certain
other cationic forms of the polysaccharide there is
evidence for Newtonian behaviour at low frequency.[10]

Figure 5 shows a plot of dynamic viscosity versus
frequency for a 1% w/v total polysaccharide preparation
(Na salt).   Four samples were measured, one with no added
salt and then three with increasing amounts of added
sodium chloride (ionic strengths of 0.01, 0.02 and 0.05).
It is evident that there is an increase in viscosity with
increasing ionic strength, with some tendency towards a
plateau.   This is not the sort of behaviour expected for
a typical polyelectrolyte where an increase in ionic
strength leads to a screening of the ionic charges (i.e.
a reduction in electrostatic repulsions) resulting in a

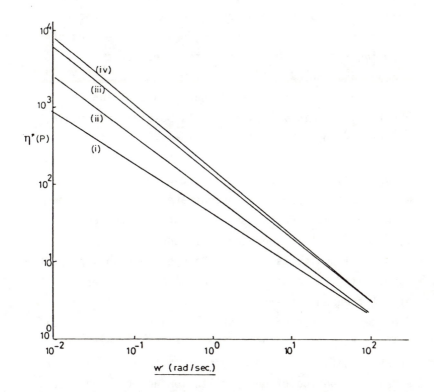

FIGURE 5   Variation in solution viscosity with ionic
            strength (I) of NaCl for the total
            polysaccharide preparation from B. polymyxa.

contraction of the coil (coil collapse) and a consequent
decrease in solution viscosity.[29]   However, it is
behaviour consistent with an association of ordered chains
with very limited (if any) flexibility.[8]   The increase
in viscosity is due to a reduction in electrostatic
repulsions between the extended, ordered molecules thereby
allowing a greater degree of intermolecular association
and aggregation.

From oscillatory measurements, in addition to
obtaining dynamic viscosity, it is possible to obtain
information on the degree of solid-like and liquid-like
nature of the material in terms of the rigidity modulus
G' and loss modulus G''.[30]   Figure 6 shows a typical
mechanical spectrum of a concentrated polysaccharide
solution (e.g. λ-carrageenan) and of a polysaccharide gel
(e.g. agar);   the frequency dependence of G', G'' and
dynamic viscosity is shown.   Polymer gels show properties
approaching those of a solid and G' predominates over G''
at all frequencies with neither showing any frequency
dependence.   For concentrated solutions of random coil
polysaccharides at high frequencies, the interchain
entanglements do not have sufficient time to come apart
within the period of one oscillation and the behaviour
is similar to true gels ($G' > G''$).   At lower frequencies
the principal response is rearrangement of the network
to accommodate the strain (flow) and G'' predominates as
for dilute solutions.

The mechanical spectrum of a 1% w/v total
polysaccharide preparation from B. polymyxa and the A-2
fraction is shown in Figure 7.   For both samples,

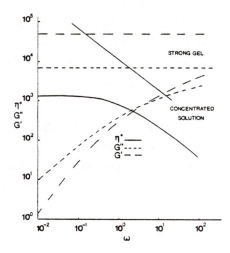

FIGURE 6   Frequency dependence of the rigidity (G') and
            viscosity (G") modulii for polysaccharide
            solutions and gels[30]

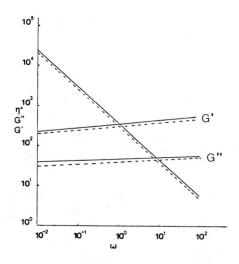

FIGURE 7   Mechanical spectrum of the total
            polysaccharide preparation from
            <u>B. polymyxa</u> (—) and the A-2 fraction (---).
            Both are 1% w/v in 50 mM NaCl.

$G' > G''$ at all frequencies. There is little frequency dependence and the spectrum is typical of that of ordered networks. The similarity between the A-2 fraction and the total polysaccharide preparation is apparent, thus emphasising the fact that it is the A-2 component which is the active one. It is not yet clear whether the slightly higher moduli of the total preparation is due to a synergistic interaction between the different sub-fractions.

ACKNOWLEDGEMENT

The authors thank Dr. D.C. Steer for providing the samples of B. polymyxa extracellular polysaccharides used in this study and for helpful discussions.

REFERENCES

1. E.R. MORRIS, A.N. CUTLER, S.B. ROSS-MURPHY and D.A. REES, Carbohydrate Polym. 1, 5-21 (1981).
2. I. SUTHERLAND, in Surface Carbohydrates of the Prokaryotic Cell, edited by I. Sutherland (Academic Press, London, 1977), Chap. 3, pp. 27-96.
3. T. HARADA, Am. Chem. Soc. Symp. Ser., 45, 265-283 (1977).
4. R. MOORHOUSE, G.T. COLEGROVE, P.A. SANDFORD, J.K. BAIRD and K.S. KANG, Am. Chem. Soc. Symp. Ser., 150, 111-124 (1981).
5. I.C.M. DEA, E.R. MORRIS, D.A. REES, E.J. WELSH, H.A. BARNES and J. PRICE, Carbohydr. Res., 57, 249-272 (1977).
6. C.A. KNUTSON, J.E. PITTSLEY and A. JEANES, Carbohydr. Res., 73, 159-168 (1979).

7. P.A. SANDFORD, Adv. Carbohydr. Chem. Biochem., 36, 265-313 (1979).
8. E.R. MORRIS, Am. Chem. Soc. Symp. Ser., 45, 81-89 (1977).
9. D.R. BAIN, private communication.
10. J.K. MADDEN, I.C.M. DEA and D.C. STEER, Carbohydrate Polym., submitted for publication.
11. N.F. DUDMAN, L.E. FRANZEN, M. McNEIL, A.G. DARVILL and P. ALBERSHEIM, Carbohydr. Res., 117, 169-183 (1983).
12. J.E. SCOTT, Methods Carbohydr. Chem., 5, 38-44 (1965).
13. A.G. WILLIAMS, J.W.T. WIMPENNY and C.J. LAWSON, Biochim. Biophys. Acta, 585, 611-619 (1979).
14. R.L. TAYLOR and H.E. CONRAD, Biochemistry, 11, 1383-1388 (1972).
15. J.H. SLONEKER and D.G. ORANTES, Nature, 194, 478-479 (1962).
16. S.I. HAKOMORI, J. Biochem. (Tokyo), 55, 205-208 (1964).
17. P.-E. JANSSON, L. KENNE, H. LIEDGREN, B. LINDBERG and J. LONNGREN, Chem. Commun. Univ. Stockholm, pp. 1-75 (1976).
18. I.H. SMITH, K.C. SYMES, C.J. LAWSON and E.R. MORRIS, Int. J. Biol. Macromol., 3, 129-134 (1981).
19. N.-X. YU, M. HISAMATSU, A. AMEMURA and T. HARADA, Carbohydrate Polym., 1, 23-30 (1981).
20. A.A. COOKE and E. PERCIVAL, Carbohydr. Res. 43, 117-132 (1975).
21. C.G.T. EVANS, R.G. YEO and D.C. ELLWOOD, in Microbial Polysaccharides and Polysaccharases, edited by R.C.W. Berkeley, G.W. Gooday and D.C. Ellwood (Academic Press, London, 1979), Chap. 3, pp. 51-68.
22. C. WHITFIELD, I.W. SUTHERLAND and R.E. CRIPPS, J. Gen. Microbiol., 124, 385-392 (1981).
23. M. HISAMATSU, I. OTT, A. AMEMURA, T. HARADA, I. NAKANISHI and K. KIMURA, J. Gen. Microbiol., 103, 375-379 (1977).
24. S.K. GHAI, M. HISAMATSU, A. AMEMURA and T. HARADA, J. Gen. Microbiol., 122, 33-40 (1981).
25. E. NINOMIYA and T. KIZAKI, Angew. Makromol. Chem., 6, 179-185 (1969).
26. M.C. CADMUS, K.A. BURTON, A.A. LAGODA and K.L. SMILEY, Bacteriol. Proc., A102 (1967).

27. M. FUMINO, Y. ICHIHARA and M. TSUJI, Jpn. Kokai Tokkyo
    Koho, 79 27,517.
28. G. ROBINSON, S.B. ROSS-MURPHY and E.R. MORRIS,
    Carbohydr. Res., 107, 17-32 (1982).
29. O. SMIDSRØD and A. HAUG, Biopolymers, 10, 1213-1227
    (1971).
30. E.R. MORRIS and S.B. ROSS-MURPHY in Techniques in
    Carbohydrate Metabolism, edited by D.H. Northcote
    (Elsevier, Amsterdam, 1981) B310.

STRUCTURE AND FUNCTION OF PLANT CELL WALL POLYSAC-
CHARIDES

A. G. DARVILL, P. ALBERSHEIM, M. MCNEIL, J. M. LAU,
W. S. YORK, T. T. STEVENSON, J. THOMAS, S. DOARES,
D. J. GOLLIN, P. CHELF, AND K. DAVIS

Department of Chemistry
Campus Box 215
University of Colorado
Boulder, CO 80309

Molecules of extraordinarily complex structure are
present in the walls that surround growing plant
cells, the so-called <u>primary</u> <u>cell</u> <u>walls</u>. These
structurally complex molecules are the polysaccha-
rides and glycoproteins that constitute more than
half the mass of primary cell walls. Present studies of primary cell walls still very much involve
elucidating the primary structures of the component
polysaccharides and glycoproteins. This paper
summarizes the current knowledge of the structures
of the primary cell wall polysaccharides that are
being characterized in our laboratory. We also
present here a summary of evidence we have obtained
in support of the hypothesis that oligosaccharide
fragments of cell wall polymers act as regulatory
molecules. This evidence includes the observation
that pure oligosaccharides can act as potent and
specific regulators of gene expression. Oligosaccharides with regulatory activities are called
<u>oligosaccharins</u>.

I. PRIMARY CELL WALLS

Virtually every cell of higher plants is encased in a cellulosic wall. The walls of growing plant cells are called primary cell walls. Primary cell walls are composed of approximately 90% complex carbohydrate and 10% protein in the form of glycoproteins[1]. In addition, there are other substituents such as methyl ethers and esters, and acetyl and feruloyl esters. Cellulose, a polymer of α-4-linked-D-glucosyl residues, constitutes 20 to 30%, and the hydroxyproline-rich glycoproteins constitute 10 to 20%, of primary cell walls[1]. The remaining polysaccharides of the cell wall are exceedingly complex molecules, none of whose structures has been completely defined. Very likely, some primary cell wall polysaccharides have yet to be discovered. Recent research has shown that the structures of many cell wall polysaccharides are so complex that even the most sophisticated technologies available today may not be capable of completely delineating the primary structure of these polymers.

Cell walls often grow ten times their length after the final division of their progenitor meristematic cell, and some cells elongate more than 100-fold. As observed under the electron microscope, the walls of these longer cells appear no thinner than those of the progenitor cells, and no physical evidence suggests that the walls of the larger cells are any weaker. Thus, there is no obvious requirement for the average size of cell wall polymers to change significantly during cell enlargement. The most apparent requirement for growth that maintains the strength and thickness of the wall is synthesis of new wall polymers and the incorporation of these polymers into the wall[1].

The primary cell wall defines not only the rate of growth of plant cells, but also the size and shape. The wall that is laid down during cell growth has great effect on the size and shape of the plant cells and, thus, on the whole plant. We recognize the tremendous diversity of species in the plant kingdom more by size and shape than by any other criteria. We can assume that constituents of the walls of the different plants must differ from species to species. The primary cell walls of all higher plants appear to contain the same general polysaccharides, sometimes in very different proportions. Furthermore, differences in fine structure do exist in the corresponding polymers of cell walls isolated from different sources.

We hypothesize that the complex carbohydrates of the walls of cells in the different organs and tissues of a single plant are also different. If this is true, then determining the structures of the primary cell walls of all the different types of plant cells will be an enormously difficult challenge, a challenge that is probably beyond our present capabilities. However, we believe that it will be possible to construct a general, consensus picture of primary cell walls without knowing the fine structural details of the walls of individual cells, although sufficient structural information is not available, at this time, to draw such a picture.

The walls of plant cells not only affect the size and shape of the cells that they surround, but also act as a barrier to the many different microbes to which plants are exposed[2,3]. Pathogens must penetrate the walls of their plant host, and, toward this end, pathogens secrete a complex mixture of cell-wall-degrading enzymes. Plants have, within their walls, proteins that can inhibit, specifically, the wall-degrading enzymes secreted by microbes[4]. The walls of plant cells also contain enzymes that can degrade the walls of microbes[5,6]. In some cases, enzymes from the walls of plant cells release, from the walls of microbes, molecules that activate a defense response in the plant[7]. Thus, when plant cell walls are exposed to attack by microorganisms, it is apparent that there is considerable interaction between molecules of plant and microbe origin. These molecular interactions determine whether a plant can be successfully invaded by a microorganism.

The walls of plants represent the bulk of all biomass. The primary cell walls of plants are a major source of dietary fiber and ultimately a source of food for human and all animals. If we are going to understand the function of primary cell walls in human and animal diet, then we must understand the chemistry of primary cell walls.

Studies of the primary structures of the cell wall polymers have been partially successful and have led to a realization that these structures are far more complex than was dreamed of just a few years ago. This enormous structural complexity led to the postulation that cell wall polysaccharides have other functions besides determining plant size and morphology. One such function was suggested, some years ago, by the discovery that oligosaccharide fragments of branched β-glucans of fungal cell walls can control gene expression in plant cells[7]. In other words, complex carbohydrates can be regulatory mole-

cules. This observation led to the insight that enzymically released fragments of the polysaccharides of primary cell walls of plants could evoke defense responses in
plants and, subsequently, to the further realization that
very specific fragments of plant cell wall polymers can
control various physiological responses of plants. These
regulatory molecules are included in the class of biologi-
cally active oligosaccharides that are called oligosaccharins. Thus, defined fragments of plant cell wall polymers
(oligosaccharins), have very specific regulatory functions
when released from covalent attachment within cell
walls. Evidence has been obtained that oligosaccharins
can regulate in plants such functions as rate of cell
growth, flowering, rooting, vegetative bud development,
and activation of mechanisms for resistance to potential
pathogens. These regulatory functions are discussed later
in this paper.

II. PARTIAL STRUCTURAL CHARACTERIZATION OF SOME CELL WALL POLYSACCHARIDES

2.1. Xyloglucan

Xyloglucan is a hemicellulose present in the cell walls of
both dicots and monocots, although in notably larger
amounts in the cell walls of dicots ($\sim$20%) and in notably
smaller amounts in the cell walls of monocots ($\sim$2%)[8]. The
structure of cell wall xyloglucan has been reviewed extensively[8,9] and, basically, it consists of a backbone of β-
4-linked-D-glucosyl residues, with D-xylosyl side chains
α-linked to 0-6 of about three out of every four glucosyl
residues. Certain xylosyl side chains have extensions of
D-Gal1$\overset{\beta}{\to}$ or L-Fuc$\overset{\alpha}{\to}$2-D-Gal1$\overset{\beta}{\to}$ at 0-2 of the xylosyl resi-

dues[8,9,10]. Occasionally, arabinosyl residues are found linked to O-2 of some of the xylosyl residues of some xyloglucans[10-14].

Considerable information has been obtained about the distribution of xyloglucans in the plant kingdom[8]. Xyloglucan has been shown to be present in the cell walls of many different dicots, including sycamore[10], soybean[15], mung bean[16,17], potato[13], runner bean[18], and tobacco[11,12,19], and monocots, including rice[20,21], bamboo[22], oats[23], and barley[24]. All of the xyloglucans are structurally very similar, but there are some differences in the side chains. Terminal L-arabinosyl residues are α-linked to O-2 of selected xylosyl residues in potato[13] and in tobacco[11]. Terminal galactosyl residues (i.e., galactosyl residues that have no fucosyl residues attached to them) are present in the xyloglocans of many cell walls[8,9,16,18,25].

Another structural feature of xyloglucans has recently been observed that indicates that this polysaccharide has greater structural complexity than originally detected. Xyloglucan isolated from the sycamore extracellular polysaccharides secreted by suspension-cultured sycamore cells has recently been shown to be substituted with O-acetyl groups (W. York, A. Darvill, M. McNeil, P. Albersheim, and A. Dell, unpublished results). This was shown by analysis of xyloglucan fragments formed by enzymic cleavage of the xyloglucan with an endo-β-1,4-glucanase. The enzymic digestion products were separated into nonasaccharide- and heptasaccharide-rich fractions. O-Acetyl substituents were identified in the nonasaccharide-, but not in the heptasaccharide-containing-fractions by [1]H-

n.m.r. and by fast—atom—bombardment mass spectrometry
(f.a.b.—m.s.). F.a.b.—m.s. indicated that the nonasaccharide fraction contained unsubstituted nonasaccharides and
nonasaccharides substituted with either one or two O—
acetyl groups. The locations on the nonasaccharide of the
O—acetyl substituents have not yet been determined.

The xyloglucan in growing walls of pea—stem cells[26]
and in suspension—cultured bean cell walls[27] has been
shown to be composed of nonasaccharides and heptasaccharides with the same structures as the nona— and heptasaccharides isolated from the xyloglucans of sycamore cell
walls and sycamore extracellular polysaccharides.

It has only been relatively recently that xyloglucans
have been clearly shown to occur in monocot cell walls[20-24]. The glucosyl residues of the monocot xyloglucans have
been shown to be substituted with xylosyl residues less
frequently than are residues of the dicot xyloglucans[27].
Monocot xyloglucans, like dicot xyloglucans, have been
shown to possess galactosyl—containing side chains[20]; the
presence in monocot xyloglucans of fucosyl—containing
side—chains has not been clearly established.

2.2. Rhamnogalacturonan I

Rhamnogalacturonan I, isolated from suspension—cultured
sycamore cell walls, is a much—studied pectic polysaccharide[28,29]. The degree of polymerization of rhamnogalacturonan I is about 2,000, and it is composed of D—galactosyluronic acid, L—rhamnosyl, D—galactosyl, L—arabinosyl,
and small amounts of L—fucosyl residues. The backbone of
rhamnogalacturonan I is composed of alternating 2—linked
L—rhamnosyl and 4—linked D—galactosyluronic acid residues[28,29,30]. The length of the polymer composed of these

strictly alternating residues is not known, but it could contain as many as 300 L-rhamnosyl and 300 D-galactosyl-uronic acid residues. Approximately half of the 2-linked L-rhamnosyl residues are branched, containing glycosyl substituents at O-4. The side chains average about seven glycosyl residues in length[28]. It has been demonstrated that seven differently linked glycosyl residues are directly attached to O-4 of the rhamnosyl residues[29]. The presence of seven differently linked glycosyl residues as the first residues of the side chains suggests that rhamnogalacturonan I possesses many different side chains.

The large number of different side chains has been confirmed by study of a mixture of the rhamnogalacturonan I side chains attached to O-4 of L-rhamnitol (backbone) residues that was prepared by treatment of rhamnogalacturonan I with lithium dissolved in ethylene diamine[31] (J. Lau, A. Darvill, M. McNeil, and P. Albersheim, unpublished results). This treatment of rhamnogalacturonan I has been shown to cleave and degrade the α-4-linked D-galactosyluronic acid residues and to convert, into rhamnitol residues, the L-rhamnosyl residues released from O-4 of the galactosyluronic acid residues. Preliminary analysis, by liquid chromatography fractionation and by fast-atom-bombardment mass spectrometry, of the resulting mixture of rhamnogalacturonan I side chains attached to rhamnitol residues analysis has shown the presence of at least 30 different side chains (J. Lau, A. Dell, M. McNeil, A. Darvill, and P. Albersheim, unpublished results). The side chains, some of which have at least 15 glycosyl residues, are rich in L-arabinosyl and D-galactosyl residues.

It is not known whether rhamnogalacturonan I is a
single polysaccharide with many different side chains
attached to a single backbone or a mixture of polysaccharides, each containing selected types of side chains
attached to a similar if not identical backbone. It is
possible that different types of plant cells produce rhamnogalacturonan I with different arrays of side chains.
More effective fractionation techniques and the ability to
analyze rhamnogalacturonan I purified from several different types of plant cells will assist in confirming one or
another of these possibilities.

Cell walls other than those of suspension-cultured
sycamore cells contain molecules very similar to rhamnogalacturonan I, although the polysaccharides have not been
studied in detail[8,32-40]. Glycosyl-linkage analysis of a
pectic polysaccharide isolated from the midrib of tobacco,
which, like sycamore, is a dicot, showed the same types of
glycosyl residues present in sycamore-cell-wall rhamnogalacturonan I, although the ratios of the various components were, in some cases, different[33]. Similar results
have been obtained when analyzing pectic polysaccharides
from potato, also a dicot[34], and from onion, a monocot[36]. A molecule with a glycosyl composition similar to
that of rhamnogalacturonan I has been isolated from rice,
another monocot[41]. Taken as a whole, the results suggest
that rhamnogalacturonan-I-like polysaccharides are present
in most if not all higher plant cell walls, although the
nature and quantity of the side chains probably varies.

2.3. Rhamnogalacturonan II

Rhamnogalacturonan II[42] is structurally very different
from rhamnogalacturonan I. Rhamnogalacturonan II is com-

pletely solubilized from suspension-cultured sycamore cell walls by the endo-α-1,4-polygalacturonase isolated from Colletotrichum lindemuthianum. Thus, it appears that rhamnogalacturonan II is covalently linked in the primary cell wall through a series of α-4-linked galactosyluronic-acid residues, perhaps through homogalacturonans. Rhamnogalacturonan II consists of approximately 60 glycosyl residues, some of which are very unusual[42]. These include 2-O-methyl fucosyl, 2-O-methyl-xylosyl, apiosyl (a branched pentosyl residue), and 3-C-carboxy-5-deoxy-L-xylosyl, the first branched-chain acidic glycosyl residue to be found in nature[43]. This new sugar was named aceric acid because it was discovered in suspension-cultured sycamore (Acer pseudoplatanus) cells. More recently, 3-deoxy-D-manno-octulosonic acid (KDO), a sugar commonly found in the core region of bacterial lipopolysaccharides and in some bacterial extracellular polysaccharides, but never before in higher plants, has been shown to be present in rhamnogalacturonan II. The KDO in this cell wall polysaccharide has a rhamnosyl residue glycosidically linked to it at O-5 (W. York, A. Darvill, M. McNeil, and P. Albersheim, submitted for publication). In addition, rhamnogalacturonan II contains a high proportion of rhamnosyl residues, but, in contrast to rhamnogalacturonan I, in which the rhamnosyl residues are 2- and 2,4-linked, the rhamnosyl residues of rhamnogalacturonan II are 3-, 3,4-, 2,3,4-, and terminally linked.

Recent studies of rhamnogalacturonan II[44] (L. Melton, M. McNeil, A. Darvill, P. Albersheim, and A. Dell, submitted) have shown that it contains two structurally complex heptasaccharides. Glycosyl-linkage analysis of rhamnogalacturonan II suggests the presence of four units of one of the heptasaccharides and two units of the other, which would account for 60% of the glycosyl residues of rhamnogalacturonan II. Evidence has also been obtained that rhamnogalacturonan II contains an α-4-linked D-octagalacturonide. The manner in which these hepta- and octasaccharides are arranged in rhamnogalacturonan II is yet to be ascertained.

Rhamnogalacturonan II has been tentatively identified in the cell walls isolated from pea, pinto bean, and tomato seedlings[42]. These identifications were based on the presence of the unusual glycosyl residues found in rhamnogalacturonan II. Using the same criteria, small amounts of rhamnogalacturonan II have been detected in the cell walls of suspension-cultured Douglas Fir, a gymnosperm (J. Thomas, A. Darvill, M. McNeil, and P. Albersheim, unpublished results), and in the cell walls of oat, a monocot[42].

III.   BIOLOGICAL ACTIVITIES OF FRAGMENTS OF PLANT CELL WALLS

Oligosaccharide fragments of plant cell wall polymers have been shown to have regulatory activities within plant tissues. Oligosaccharides with regulatory properties are called oligosaccharins. Investigating the regulatory effects of oligosaccharins is a relatively new area of research. Therefore, it is not known how many biological functions can be controlled by oligosaccharins. Several such regulatory molecules will be described in this section.

3.1. Cell Wall Fragments as Elicitors of Phytoalexins

The synthesis of phytoalexins (antibiotics) is considered a general defense mechanism of plants[45]. Fragments of homogalacturonan have been shown to induce plant tissue to synthesize phytoalexins. Molecules that induce the synthesis of phytoalexins, including structurally specific oligo-β-glucosides from fungal cell walls[45-47] and homogalacturonan fragments from plant cell walls, are termed elicitors. Elicitors change the metabolism of receptive plant cells so that the mRNAs and enzymes responsible for phytoalexin synthesis are themselves synthesized de novo[48]. Exactly how elicitors activate the expression of specific genes is not understood[48].

Homogalacturonan and other galactosyluronic acid-rich fragments have been shown to be endogenous elicitors of phytoalexins by two separate research groups working with two different plant systems. These elicitors are termed endogenous elicitors because they are not active constitutively, but instead have to be released from covalent attachment within the cell wall to be effective as elicitors. In one of these systems, based on a soybean cotyledon bioassay for elicitor activity, homogalacturonan fragments were produced from cell walls in two different ways[49-51]. The first method of preparation involved partial acid hydrolysis of soybean cell walls[49,50]. The resulting mixture of oligosaccharides was purified by ion-exchange and gel-filtration chromatography. The greatest elicitor activity was associated with the α-4-linked-D-dodecagalacturonide-containing fraction, although elicitor activity was observed in fractions containing oligogalacturonides from 10 to 13 residues long[49]. Chemical analyses showed that the dodecagalacturonide-containing fraction was composed of 99% galactosyluronic acid. Negative-ion fast-atom-bombardment mass spectrometry exhibited a (M - H) ion at m/z 2129, which determined that the dominant oligosaccharide was composed of exactly 12 galactosyluronic acid residues. ^{1}H-n.m.r. showed that the anomeric linkages of the galactosyluronic acid residues were α-, and glycosyl-linkage analysis showed that the galactosyluronic acid residues were 4-linked. The elicitor activity of the α-4-linked-D-dodecagalacturonide was lost by incubation with a pure endo-α-1,4-polygalacturonase, which established that contiguous α-4-linked-D-galactosyluronic acid residues are needed for elicitor activity. The same elicitor-active oligosaccharin was obtained after partial acid hydrolysis of citrus pectin[49].

A second elicitor of phytoalexins was identified and purified after incubation of various substrates with an endo-α-1,4-polygalacturonic-acid lyase (PGA lyase)[51] (K. Davis, A. Darvill, and P. Albersheim, unpublished results). This enzyme cleaves unesterified regions of homogalacturonans by an elimination reaction yielding oligogalacturonans containing a 4,5-unsaturated nonreducing terminal D-galactosyluronic acid residue. The enzyme was purified to homogeneity from the bacterium Erwinia carotovora[51]. The purified PGA lyase applied to soybean cotyledons elicits phytoalexins. Further, the enzyme produces elicitor-active galactosyluronic-acid-rich fragments (oligosaccharins) from soybean cell walls, polygalacturonic acid, and citrus pectin.

The most elicitor-active fragment produced by PGA-lyase treatment of polygalacturonic acid has been purified by ion-exchange and gel-filtration chromatography. This molecule was characterized by fast-atom-bombardment mass spectrometry and by other chemical methods and was shown to be an α-4-linked decagalacturonide with a 4,5 unsaturated nonreducing terminal residue (K. Davis, A. Darvill, A. Dell, and P. Albersheim, unpublished results). This molecule exhibited half-maximal elicitor activity at approximately 2 μg/cotyledon.

West and co-workers[52-54] have purified an enzyme, from the fungal pathogen Rhizopus stolonifer, that elicits phytoalexins in its host, castor bean (Ricinus communis). They purified this enzyme to homogeneity and have shown it to be an endo-α-1,4-polygalacturonase. They used the pure enzyme to release oligogalacturonides from a cell-free particulate fraction (cell walls) of castor bean. They found that an α-4-linked tridecagalacturonide was the most active oligogalacturonide elicitor (oligosaccharin) in their system. The differences in the optimum size of the oligogalacturonides--10, 12, and 13 residues-- in the different systems examined may reflect differences in the activities of oligogalacturonide-degrading enzymes in the plant tissues being elicited. The available evidence suggests that the release of endogenous oligogalacturonide-containing elicitors by microbes is a widespread, physiologically important defense mechanism for plants. Of interest in this area is that the endogenous oligogalacturonide elicitor and the fungal derived hepta-β-glucoside elicitor exhibit a pronounced synergism, that is, about 50-fold less hepta-β-glucoside is required when the two elicitors are applied simultaneously to soybean cotyledons[45]. A more detailed review of the significance of the endogenous elicitors has recently been published[45].

3.2. Proteinase Inhibitor Inducing Factor

Fragments of pectic polysaccharides have been shown to be effective in another process by which plants protect themselves from potential pathogenic microbes and from insects. Ryan and co-workers have shown that mechanical injury will induce the systemic synthesis in plants of large amounts of proteins that inhibit insect and microbial proteinases[55,56]. Presumably, the proteinase inhibitors protect the plants by making them less digestible to invading insects and microbes. The signal by which tissues distant from the injury are induced to respond and synthesize proteinase inhibitors has been named PIIF (Proteinase Inhibitor Inducing Factor)[56].

It has been found that fragments of pectic polysaccharides (oligosaccharins) have PIIF activity[57,58]. A highly purified preparation of the sycamore cell wall pectic polysaccharide rhamnogalacturonan I exhibited PIIF activity in tomato seedlings[57]. More recent studies by Ryan and co-workers have suggested that the PIIF-active substance extracted from wounded tomato leaves may be predominantly composed of α-4-linked galactosyluronic acid residues, although the most active fragments are considerably smaller than the phytoalexin-elicitor-active fragments[58].

3.3. Cell Wall Fragments That Kill Plant Cells

Another cell wall fragment that may be involved in the processes by which plants protect themselves from potential pathogens has been recently detected. This cell wall fragment causes death of plant cells. An apparent sacrifice of host cells, known as hypersensitive cell death, is an early reflection of a widely observed and very important defense response of plants against microbes[59]. Hypersensitive cell death somehow results in the slowing down of the invading organism's growth, providing time for other defense reactions of the plants to stop the attempted infection. Death-inducing cell wall fragments (oligosaccharins) were prepared by partial acid hydrolysis of the purified primary walls of suspension-cultured sycamore[60] and maize cells (S. Doares, A. Darvill, and P. Albersheim, unpublished results). Sub-lethal doses of the toxic fragments were assayed by their inhibitory effect on uptake and incorporation into protein of [^{14}C]leucine by cultured sycamore and maize cells. The measure of protein synthesis was considered a measure of cell vitality[60]. The cell wall fragments from both sycamore (a dicot) and maize (a monocot) actively inhibit protein synthesis in both sycamore and maize cells (S. Doares, A. Darvill, and P. Albersheim, unpublished results).

The toxic fragments, which may be pectic in composition, might account for the well-known phytotoxicity of pectic enzymes[61-63]. This view is given circumstantial support by the finding that plasmolysis protects plant tissues from the toxicity of both pectic enzymes[64,65] and the fragments[60]. It is interesting to hypothesize that invading microbes cause hypersensitive death of plant cells by enzymically solubilizing toxic pectic fragments (oligosaccharins) from plant cell walls that surround the hypersensitive plant cells. Preliminary evidence suggests that the rice pathogen Pyricularia oryzae secretes heat-labile molecules (putative enzymes) that can release from cell walls oligosaccharides that can inhibit protein synthesis (S. Doares, A. Darvill, and P. Albersheim, unpublished results).

3.4. Inhibition of 2,4-D-Induced Growth by a Fragment Isolated from Xyloglucan

An oligosaccharin fragment of xyloglucan has been shown to inhibit auxin (2,4-D)-induced pea epicotyl growth[66]. The fragment was obtained from xyloglucan secreted by suspension-cultured sycamore cells by treatment with endo-β-1,4-glucanase. The resulting oligosaccharide products were separated by P-2 gel chromatography. Two major fractions were obtained, the first, rich in a nonasaccharide of known structure, and the second, rich in a heptasaccharide also of known structure[10]. The nonasaccharide fraction differs from the heptasaccharide fraction by possessing an L-fucosyl $1\overset{\alpha}{\rightarrow}2$-D-galactosyl disaccharide linked to O-2 of a xylosyl side chain of the heptasaccharide. The nonasaccharide fraction, at a concentration of about 10^{-8} M inhibits 70% to 90% of the 2,4-D-stimulated growth of pea epicotyls. The xyloglucan heptasaccharide fraction showed no significant inhibitory activity at similar concentrations. The inhibitory activity of the nonasaccharide fraction exhibited a sharp concentration optimum, ruling out the possibility that the nonasaccharide fraction was toxic to the pea tissue. The nonasaccharide fraction inhibited the 2,4-D stimulated growth at a concentration that was approximately 1% of the molar concentration of 2,4-D necessary for maximal promotion of elongation growth[66].

3.5. Inhibition of Flowering of Lemna gibba G3 (Duckweed) by a Plant Cell Wall Fragment

The control of flowering in plants is a complex process that is controlled by many environmental factors including day length, plant maturity, and stress. It has been sug-

gested that, at the molecular level, both flower-promoting
and flower-inhibiting factors participate in the control
of flowering[67]. None of these factors, however, has been
identified. Partial acid hydrolysis releases, from the
walls of suspension-cultured sycamore cells, a cell wall
fragment (oligosaccharin) that inhibits the flowering of
Lemna gibba G3 and stimulates the rate of growth of the
vegetative fronds. This observation suggested that the
fragments were not merely toxic, but were actually switch-
ing the destination of developing cells from flowering to
vegetative. The identity of the flowering-inhibiting cell
wall fragments has not been established, although evidence
has been shown that the active fragments are pectic in
nature[68]. It is not known whether these fragments will
affect the flowering of other plant species. This work
does indicate, however, that cell wall fragments should be
given serious consideration as molecules that may play a
role in the control of flowering.

3.6. The Control of Morphogenesis in the Thin Cell-Layer
Explants of the Flowering Branches of Tobacco

Aquiring the ability to regulate morphogenesis in plants
has long been an elusive goal. Varying the pH of culture
media and the ratios of growth regulators has yielded
evidence that tissue cultures of some plant species can be
influenced to form vegetative buds, roots or a combination
of these in vitro. One such system, thin cell-layer
explants of tobacco, has been established by Dr. Kiem Tran
Thanh Van and her research group at the Phytotron Laboratory at Gif sur Yvette, France[69]. Over a ten-year period
they have established the ability to control the morphogenesis of 1 x 10 mm-thin cell-layer strips (the epidermis

plus one layer of endodermis and about three layers of parenchyma) taken from the basipetal portions of the flower branches of tobacco. The morphogenesis of the system is controlled by the pH and auxin concentration of the liquid medium on which the explants are grown. The assay involves growing 20 explants for 20 days in a Petri dish. All of the explants in each dish can be induced to form callus, vegetative buds, roots, or flowers.

In collaboration with the Gif group, we have been testing the effects of cell wall oligosaccharide fragments on the morphogenesis of the tobacco thin-cell-layer explants. This system is ideal for examining the effect of oligosaccharide mixtures on the morphogenesis of plant tissues. The bioassay is sterile, and most or all of the cells of the explants are in contact with the incubation media and, therefore, with the exogeneously supplied oligosaccharides. We supplied the group in Gif with several mixtures of cell wall fragments. The two oligosaccharide mixtures that have been studied most thoroughly were produced by partial base solubilization of cell walls and by endopolygalacturonase solubilization of cell walls. These mixtures of cell wall fragments were obtained by treating purified cell walls isolated from suspension-cultured sycamore cells.

Explants grown under conditions that induce the formation of only flowers develop floral shoots and vegetative buds instead when base-solubilized fragments are added to the culture media. Adding base-solubilized fragments to explants that, in the absence of fragments form only vegetative buds, inhibited formation of vegetative buds and induced the formation of roots.

In similar experiments, adding endopolygalacturonase-solubilized wall fragments to explants that would form only flowers in the absence of added fragments inhibited flower formation and induced the explants to form vegetative buds. Adding endopolygalacturonase-solubilized cell wall fragments also induced formation of vegetative buds in explants that otherwise formed only callus. Adding endopolygalacturonase-solubilized fragments to explants that normally would form only vegetative buds, inhibited the formation of the vegetative buds and induced formation of flowers.

In these experiments, the mixtures of wall fragments were added at a concentration of 1 and/or 10 µg/ml. We estimate the concentration of active oligosaccharins in the fragment mixtures to be about 10^{-8} M. This calculation matches the specific activity observed for other oligosaccharins[45]. Also, like other oligosaccharins, the fragments exhibited activity over a rather narrow (one or two orders of magnitude) range of concentration; the morphogenetic effects disappeared when the concentration was too high or too low. It should be emphasized that adding cell wall fragments had no effect on the pH of the media in which the explants were incubated. Furthermore, the culture media contained 30 mg/ml of glucose, so the effect of the wall fragments is neither a non-specific carbohydrate effect nor an energy effect.

IV. CONCLUSION

The observation that cell-wall-derived oligosaccharins can be active as regulatory molecules suggests that such oligosaccharins should be considered as a class of molecules active in the control of growth, differentiation, and defense responses in plants, and, further, that these molecules may be in situ regulators of morphogenesis.

ACKNOWLEDGMENTS

This research was supported by U. S. Department of Energy grant nos. DE-ACO2-76ERO-1426 and DE-ACO2-84ER13161. The authors thank Anne Dell, Imperial College of Science and Technology, London for all f.a.b.-m.s analyses; and Kiem Tran Thanh Van, Patrick Toubart, and Alain Cousson for all the tobacco thin cell-layer explant experiments; and Leigh Kirkland for editorial assistance.

REFERENCES

1. M. MCNEIL, A. G. DARVILL, STEPHEN C. FRY, AND P. ALBERSHEIM, Ann. Rev. Biochem., 53: 625-663, (1984).
2. D. F. BATEMAN, AND H. G. BASHAM, in Encyclopedia of Plant Physiology. Physiological Plant Pathology, New Series, edited by R. Heitefuss, and P. H. Williams, (Springer-Verlag, Berlin, 1976).
3. P. ALBERSHEIM, AND A. ANDERSON-PROUTY, Ann. Rev. Plant Physiol., 26, 31-52, (1975).
4. P. ALBERSHEIM, AND A. J. ANDERSON, Proc. Natl. Acad. Sci. USA, 68, 1815-1819, (1971).
5. K. CLINE, AND P. ALBERSHEIM, Plant Physiol., 68, 207-220, (1981).
6. K. CLINE, AND P. ALBERSHEIM, Plant Physiol., 68, 221-228, (1981).
7. A. AYERS, J. EBEL, B. VALENT, AND P. ALBERSHEIM, Plant Physiol., 57, 751-759, (1981).
8. A. G. DARVILL, M. MCNEIL, P. ALBERSHEIM, AND D. P. DELMER, in The Biochemistry of Plants, edited by N. E. Tolbert, (Academic Press, New York, 1980), 1, 91-162.
9. M. MCNEIL, A. G. DARVILL, AND P. ALBERSHEIM, in Progress in the Chemistry of Organic Natural Products, edited by W. Herz, H. Grisebach, and G. W. Kirby, (Springer-Verlag, New York, 1979), 37, 191-249.
10. W. D. BAUER, K. TALMADGE, K. KEEGSTRA, AND P. ALBERSHEIM, Plant Physiol., 51, 174-187, (1973).

11. S. EDA, AND K. KATO, Agric. Biol. Chem., 42(2), 351–357, (1978).

12. M. MORI, S. EDA, AND K. KATO, Carbohydr. Res., 84, 125–135, (1980).

13. S. G. RING, AND R. R. SELVENDRAN, Phytochemistry, 20, 2511–2519, (1981).

14. S. EDA, H. KODAMA, Y. AKIYAMA, M. MORI, K. KATO, A. ISHIZU, AND J. NAKANO, Agric. Biol. Chem., 47, 1791–1797, (1983).

15. T. HAYASHI, Y. KATO, AND K. MATSUDA, Plant and Cell Physiol., 21, 1405–1418, (1980).

16. Y. KATO, AND K. MATSUDA, Agric. Biol. Chem., 44, 1751–1758 (1980).

17. Y. KATO, AND K. MATSUDA, Agric. Biol. Chem., 44, 1759–1766 (1980).

18. M. A. O'NEILL, AND R. R. SELVENDRAN, Carbohydr. Res., 111, 239–255 (1983).

19. M. MORI, S. EDA, AND K. KATO, Agric. Biol. Chem., 43, 145–149 (1979).

20. N. SHIBUYA, AND A. MISAKI, Agric. Biol. Chem., 42, 2267, (1978).

21. Y. KATO, S. ITO, K. IKI, AND K. MATSUDA, Plant and Cell Physiol., 23, 351–364, (1982).

22. Y. KATO, R. SHIOZAWA, S. TAKEDA, S. ITO, AND K. MATSUDA, Carbohydr. Res., 109, 233–248, (1982).

23. J. M. LABAVITCH, AND P. M. RAY, Phytochemistry, 17, 933–937, (1978).

24. Y. KATO, K. IKI, AND K. MATSUDA, Agric. Biol. Chem., 45, 2745–2753, (1981).

25. Y. KATO, T. KOYAMA, T. HAYASHI, AND K. MATSUDA, Proc. Intl. Symp. on Glycoconjugates, 7th, III, 37, (1983).

26. T. HAYASHI, AND G. A. MACLACHLAN, Plant Physiol. Suppl, 72, 59, (1983).

27. B. M. WILDER, AND P. ALBERSHEIM, Plant Physiol., 51, 889–893, (1973).

28. M. MCNEIL, A. G. DARVILL, AND P. ALBERSHEIM, Plant Physiol., 66, 1128–1134, (1980).

29. M. MCNEIL, A. G. DARVILL, AND P. ALBERSHEIM, Plant Physiol., 70, 1586–1591, (1982).

30. J. M. LAU, M. MCNEIL, A. G. DARVILL, AND P. ALBERSHEIM, Carbohydr. Res., in press, (1984).

31. A. J. MORT, AND W. D. BAUER, J. Biol. Chem., 257, 1870–1875, (1982).

32. H. KONNO, AND Y. YAMASAKI, Plant Physiol., 69, 864–868, (1982).

33. S. EDA, AND K. KATO, Agric. Biol. Chem., 44, 2793–2801, (1980).

34. S. ISHII, Phytochemistry, 20, 2329-2333, (1981).
35. A. J. BARRETT, AND D. H. NORTHCOTE, Biochem. J., 94, 617-627, (1965).
36. S. ISHII, Phytochemistry, 21, 778-780, (1982).
37. R. W. STODDART, A. J. BARRETT, AND D. H. NORTHCOTE, Biochem. J., 102, 194-204, (1967).
38. J. A. DE VRIES, C. H. DEN UIJL, A. G. J. VORAGEN, F. M. ROMBOUTS, AND W. PILNIK, Carbohydr. Polymers, 3, 193-205, (1983).
39. J. A. DE VRIES, A. G. J. VORAGEN, F. M. ROMBOUTS, AND W. PILNIK, Carbohydr. Polymers, 1, 117-127, (1981).
40. J. A. DE VRIES, F. M. ROMBOUTS, A. G. J. VORAGEN, AND W. PILNIK, Carbohydr. Polymers, 2, 25-33, (1982).
41. N. SHIBUYA, AND T. IWASAKI, Agric. Biol. Chem., 42, 2259-2266, (1978).
42. A. G. DARVILL, M. MCNEIL, AND P. ALBERSHEIM, Plant Physiol., 62, 418-422, (1978).
43. M. W. SPELLMAN, M. MCNEIL, A. G. DARVILL, AND P. ALBERSHEIM, Carbohydr. Res., 122, 115-129, (1983).
44. M. W. SPELLMAN, M. MCNEIL, A. G. DARVILL, AND P. ALBERSHEIM, Carbohydr. Res., 122, 131-153, (1983).
45. A. G. DARVILL, AND P. ALBERSHEIM, Ann. Rev. Plant Physiol., 35, 243-275, (1984).
46. A. R. AYERS, J. EBEL, B. S. VALENT, AND P. ALBERSHEIM, Plant Physiol., 57, 760-765, (1976).
47. A. R. AYERS, J. EBEL, F. FINELLI, N. BERGER, AND P. ALBERSHEIM, Plant Physiol., 57, 751-759, (1976).
48. J. A. BAILEY, in Phytoalexins, edited by J. A. Bailey, and J. W. Mansfield, (Halsted Press, John Wiley and Sons, New York, 1982), pp. 289-318.
49. E. A. NOTHNAGEL, M. MCNEIL, P. ALBERSHEIM, AND A. DELL, Plant Physiol., 71, 916-926, (1983).
50. M. G. HAHN, A. G. DARVILL, AND P. ALBERSHEIM, Plant Physiol., 68, 1161-1169, (1981).
51. K. R. DAVIS, G. D. LYON, A. G. DARVILL, AND P. ALBERSHEIM, Plant Physiol. Suppl. 69, 142, (1982).
52. R. J. BRUCE, AND C. A. WEST, Plant Physiol., 69, 1181-1188, (1982).
53. S.-C. LEE, AND C. A. WEST, Plant Physiol., 67, 633-639, (1981).
54. S.-C. LEE, AND C. A. WEST, Plant Physiol., 67, 640-645, (1981).
55. C. A. RYAN, TIBS, 7, 148-150, (1978).
56. T. R. GREEN, AND C. A. RYAN, Science, 175, 776-777, (1972).

57. C. A. RYAN, P. BISHOP, G. PEARCE, A. G. DARVILL, M. MCNEIL, AND P. ALBERSHEIM, Plant Physiol., 68, 616–618, (1981).
58. P. D. BISHOP, D. J. MAKUS, G. PEARCE, AND C. A. RYAN, Proc. Natl. Acad. Sci. USA, 78, 3536–3540, (1981).
59. D. J. MACLEAN, J. A. SARGEMT, I. C. TOMMERUP, AND D. S. INGRAM, Nature, 249, 186–187, (1974).
60. N. YAMAZAKI, S. C. FRY, A. G. DARVILL, AND P. ALBERSHEIM, Plant Physiol., 72, 864–869, (1983).
61. D. F. BATEMAN, in Biochemical Aspects of Plant-Parasite Relationships, edited by J. Friend, and D. R. Threlfall, (Academic Press, New York, 1976), pp. 79–103.
62. E. C. HISLOP, J. P. R. KEON, AND A. H. FIELDING, Physiol. Plant Pathol., 14, 371–381, (1979).
63. H. T. TRIBE, Ann. Bot. NS, 19, 351–368, (1955).
64. H. G. BASHAM, AND D. F. BATEMAN, Phytopathology, 65, 141–153, (1975).
65. I. TAKEBE, Y. OTSUKI, AND S. AKIO, Plant Cell Physiol., 9, 115–124, (1968).
66. W. S. YORK, A. G. DARVILL, AND P. ALBERSHEIM, Plant Physiol., 75, 295–297, (1984).
67. G. BERNIER, J.-M. KINET, AND R. M. SACHS, The Physiology of Flowering (CRC Press, Inc., Boca Raton, 1981), Vols. 1 (149 pp.), and 2 (231 pp.).
68. D. J. GOLLIN, A. G. DARVILL, AND P. ALBERSHEIM, Biol. of the Cell, (Biologie Cellulaire), 51(2), in press, (1984).
69. K. M. TRAN THANH VAN, Ann. Rev. Plant Physiol. 32, 291–311, (1981).

INTRODUCTION TO ROUND TABLE DISCUSSION

G. O. ASPINALL
Department of Chemistry
York University
Downsview, Toronto, Ontario M3J 1P3, Canada

I am happy to introduce this Round Table Discussion on new
carbohydrate polymers. Recent developments include both
the discovery of new carbohydrate polymers of interest
because of their unusual bulk properties, and the recognition that already known types of polysaccharide, in themselves or as fragments formed from them, have information
carrying potential of biological importance. These developments add excitement to a field which for too long has
maintained an image of rather uninteresting, literally
chemically monotonous, macromolecules.

The important physical properties of cellulose[1], biologi-
cally and in industrial utilization, depend on regularities,
not only of covalent structure but also of chain conformation and inter-chain hydrogen bonding, so that highly regular chain alignments result in the stabilization of
bundles of cellulose chains in the microfibril. In contrast to cellulose and other homoglycans of uniform linkage type, most of the new carbohydrate polymers are more
complex in chemical constitution and it will be useful to
consider three types of structural variation in polysaccharides that form the basis for the correlation of structure with physical properties and/or biological function
(see Figure). Polysaccharides of the first type, mainly

Regular repetition, e.g. bacterial O-antigens

Masked repetition, e.g. glycosaminoglycans

Regular backbone but irregular side-chains and/or

irregular insertions

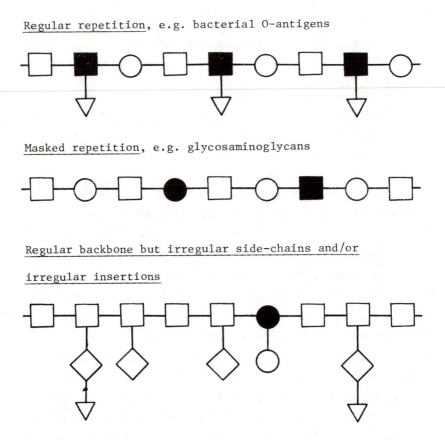

FIGURE   Variable structures in polysaccharides

of bacterial origin[2], are those of highly regular repetitive structure and are exemplified by the capsular polysaccharides (K antigens) of Gram-negative bacteria, which
have been discussed by Guy Dutton, and by the O antigen
components of the lipopolysaccharides (LPS) from other
Gram-negative bacteria. Their structures are highly

regular and specific for any given strain or serotype, but
may vary markedly from one organism to another. The high
degree of structural regularity is a direct consequence of
the biosynthetic process in which the oligosaccharide repeating unit is assembled in the cell by covalent attachment to a lipid carrier and then transported across the
membrane for polymerization on the outer surface and presentation to the external environment.

The second group of polysaccharides comprises those with
masked repetition in which the regularity of structure
dictated by biosynthesis has been modified and at first
sight obscured by post-polymerization changes. Alginic
acid provides an example of a modified homoglycan in which
some of the original 4-linked β-$\underline{D}$-mannuronic acid residues
have undergone epimerization in the polysaccharide chain
to those of α-$\underline{L}$-guluronic acid. The properties of alginic
acid samples from different sources are dependent on the
extent of such changes and on the relative distribution of
modified and unmodified units in the chain in terms of
mannuronic acid blocks, guluronic acid blocks, and regions
of mixed type[3].

Most of the examples in this second group of polysaccharides are those synthesized in a regularly alternating
sequence of A and B units, but in which some units of one
or both types have been modified subsequent to polymerization. Such polysaccharides of commercial interest by virtue of their bulk properties are those from red algae,
notably the carrageenans and agarose and related substances.
As shown in the extensive and now classical investigations
of Dai Rees and his collaborators[4], the gelling properties
of these polysaccharides may be considered in terms of

regions of regular repetitive structure that permit the
formation of junction zones with chains in close association, but with periodic departures from regularity where
complete alignment of chains is no longer possible.   The
separated junction zones joined through non-associating
chains thus give rise to an open network structure in the
interstices of which solvent molecules are entangled.
The glycosaminoglycan components of proteoglycans provide
another set of examples of polysaccharides with modified
repeating AB type structures[5].   The gross physical properties of these polysaccharides resulting from macromolecular
interactions, including self-association, are doubtless
dependent on the extent and distribution of modified sugar
residues in the overall structures.   In addition, in some
of these polysaccharides, certain modified regions may
have very precise structures responsible for biological
specificity, as exemplified by those regions in heparin
that possess high binding activity towards antithrombin III.
Benito Casu will discuss aspects of these properties of
heparin later in this Workshop.
I turn last to a third group of polysaccharides, mainly of
plant origin, where concepts of regularity and irregularity
are less clearly defined[6].   Most of these polysaccharides
contain regular backbones and to a first approximation may
be described solely in these terms.   However, the majority
of these substances are now known to be heteropolysaccharides where deviations from regularity of structure involve
the attachment of side-chains of variable length and/or
the insertion in the main chain of different sugar residues
or of the same sugars in different linkage type.   In large
measure adequate methods to determine the regularity with

which these additional structural features occur are not
yet available. Nevertheless we do know that substantial
segments of structure are regular homoglycan chains inter-
rupted only occasionally by the insertion of a different
sugar residue, e.g. an L-rhamnose residue in the galacturonan chains of pectins and related polysaccharides, or a
residue of the same sugar in different linkage type, e.g.
a 6-linked β-D-galactopyranose residue between blocks of
3-linked β-D-galactopyranose residues in the internal
chains of arabinogalactans. Other apparent irregularities
arise through varying degrees of chain extension of side-
chains containing internal units of the same linkage type.
These as yet incompletely defined departures from strictly
regular structures may be entirely necessary for the development of certain physico-chemical properties and for the
performance of biological function. In addition, as we
have heard in the very exciting new proposals from Peter
Albersheim and his colleagues, plant cell wall polysaccharides need not be restricted to an essentially static role
in nature, but may, under very specific enzymic degradation,
generate oligosaccharides of precise structure and defined
biological function.

In summary, the potential applications of new carbohydrate
polymers will be those which take advantage of bulk physical properties or of those features with biological specificity. Insofar as these properties are based ultimately
on chemical constitution and stereochemistry, account
should be taken, not only of those which arise from regularities in structure, but also of those different characteristics which result from departures from chemically
monotonous structures.

REFERENCES

1. R. H. MARCHESSAULT and P. R. SUNDARARAJAN, in The
    Polysaccharides, Vol. 2, edited by G. O. Aspinall
    (Academic Press, New York, 1983), Chap. 2, pp. 11-95.
2. L. KENNE and B. LINDBERG, in The Polysaccharides,
    Vol. 2, edited by G. O. Aspinall (Academic Press,
    New York, 1983), Chap. 5, pp. 287-363.
3. T. J. PAINTER, in The Polysaccharides, Vol. 2,
    edited by G. O. Aspinall (Academic Press, New York,
    1983), Chap. 4, pp. 195-285.
4. D. A. REES, E. R. MORRIS, D. THOM, and J. K. MADDEN,
    in The Polysaccharides, Vol. 1, edited by G. O.
    Aspinall (Academic Press, New York, 1982), Chap. 5,
    pp. 195-290.
5. L.-Å. FRANSSON, in The Polysaccharides, Vol. 3,
    edited by G. O. Aspinall (Academic Press, New York),
    in press.
6. A. M. STEPHEN, in The Polysaccharides, Vol. 2,
    edited by G. O. Aspinall (Academic Press, New York,
    1983), Chap. 3, pp. 97-193.

CELLULOSE AND AMYLOSE: CRYSTAL STRUCTURES AND PROPERTIES

ANATOLE SARKO
Department of Chemistry
State University of New York
College of Environmental Science & Forestry
Syracuse, New York 13210 USA

Both cellulose and amylose, in addition to being naturally crystalline, are able to exist in many different crystalline forms. The study of these polymorphs by the newer, computer-assisted diffraction analysis methods has resulted in detailed structural characterization of both polymers. The accumulated data on the conformations and physical characteristics of the two polysaccharides are providing a structural basis for a better understanding of their solution and solid state properties, mechanisms of interconversions, biosynthesis, and morphology.

1. INTRODUCTION

Among the many polysaccharides of plant, aquatic, and microbial systems, cellulose and starch occupy a unique position as the world's most widely utilized natural products. While cellulose forms the basis of large scale fiber, film, and chemical derivatives industries, starch has a commanding position in the food industry. The chemistry of both polysaccharides has been studied for many decades. However, it is only relatively recently that the physical structure and

morphology of these polymers have been studied more inten-
sively, in the realization of their importance in determining properties. In their native states, both cellulose and
starch are substantially -- although not completely -- crystalline which facilitates their study by diffraction methods. At the same time, their incomplete degree of crystallinity creates a complicated and not easily characterized
morphology. But as will be shown later, diffraction analysis can be of help in such characterization also. The crystallinity of both native polysaccharides extends to their
many "man-made" forms, or polymorphs, as well as derivatives
and complexes. Both substances thus present large families
of crystalline structures ideally suited to study by diffraction methods. The result has been the accumulation of a
significant body of knowledge on the conformational, crystalline, and morphological structural aspects of both polysaccharides. This knowledge is proving of considerable value in establishing a structural basis for the properties of
both materials.

It is the intent here to review the more recent results
of modern diffraction analysis applied to cellulose and amylose. Because these results have been obtained through the
development and continual refinement of new, computer-based
analysis methods, the latter will be described briefly.

2. POLYMER DIFFRACTION ANALYSIS

Although the majority of polymer crystallographic characterization studies have utilized x-ray diffraction techniques,
electron diffraction has of late become increasingly more
useful, particularly when applied in conjunction with x-ray
diffraction. In either case, polymer crystallography dif-

fers considerably from the classical, single crystal analysis of small molecules. The reasons for this departure are readily apparent in the x-ray fiber diagram of native cellulose, shown in Fig. 1. The polycrystalline morphology combined with a small crystallite size effectively reduce the available diffraction data set to below the limit where direct methods can lead to the final structure solution. Instead, a modeling approach must be used in which an initial model of the structure is established by any appropriate method, followed by its refinement using diffraction intensities and stereochemical principles. This procedure has now been successfully applied to a large number of polysaccharides for which x-ray fiber diagrams can be obtained (see, for example, refs. 1 and 2). Considerably more difficulty is encountered when only powder x-ray diffraction diagrams are available, as in the case of starch. In the latter cases, more involved sample preparation and analysis methods are necessary, as will be shown later.

In principle, the model-based diffraction analysis applies equally well to all diffraction methods -- x-ray, electron, neutron, etc. Its application is schematically illustrated in Fig. 2. From the initial model, a set of calculated structure factor amplitudes ($\underline{F}_{calc}$) is obtained and compared with the corresponding observed set ($\underline{F}_{obs}$), obtained from the measured diffraction intensities. The differences between the two sets of quantities are used to alter the initial model. This process is repeated until the calculated and observed structure factor amplitudes agree, at which time the structure can be considered solved. It should be added that during this iterative process, the stereochemistry of the model must be continually monitored

A. SARKO

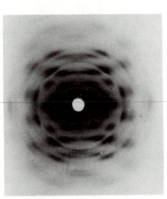

FIGURE 1   X-ray fiber diagram of
            ramie cellulose. (Fiber
            axis is vertical).

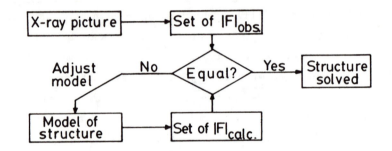

FIGURE 2   Flow diagram of structure refinement in diffrac-
            tion analysis.

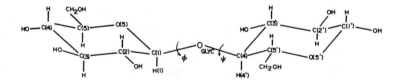

FIGURE 3 ϕ, ψ rotations in polysaccharides.

to guard against altering the model in structurally unreasonable directions.

There are three key elements involved in this procedure: (1) the establishing of the initial model, (2) its stereochemical refinement, and (3) the final refinement with structure factors, leading to the structure solution.

2.1. The Initial Model

The initial model can be established in a number of ways, but the following two methods are the most useful. In both cases, a polysaccharide chain is created by linking together the appropriate carbohydrate units. The atomic coordinates of the latter are generally taken from single crystal carbohydrate structure analyses. The major features of the polysaccharide chain can be altered either by performing rotations about the bonds linking successive residues (the ϕ, ψ rotations, cf. Fig. 3), or by rotating the residues about the virtual bond (VB), a vector connecting successive glycosidic oxygens (cf. Fig. 4). The first procedure is most useful when the diffraction diagram does not clearly indicate a suitable starting model. Conversely, when the diffraction diagram unequivocally specifies the major characteristics of the initial chain model (i.e., yields $\underline{n}$, the number of sugar residues per turn of helix, and $\underline{h}$, the rise per residue along the helix axis), the second procedure is more useful. However, in both cases, the conformational energy of each model must be calculated in order to determine its stereochemical feasibility. A number of energy expressions are available to perform such calculations, viz., Lennard-Jones, Buckingham, Kitaigorodskii, etc. (3) When, for example, this is done for the ϕ, ψ procedure and the energy corresponding to each combination of angles is plotted as a function of the latter, an energy contour plot (Ramachandran plot) is obtained. The plot shows the regions of energetically allowed and forbidden models, as well as energy minima which are likely to contain the most probable models. Such a plot for cellulose is shown in Fig. 5. On the same plot, contours of n and h can be superimposed, allowing an immediate determination of the helical parameters of any desired model.

The virtual bond rotation procedure is generally much simpler, as the n - h parameters are fixed and the only major conformational variable changing during the rotation is the bond angle θ at the glycosidic oxygen (cf. Figs. 5 and 6). A suitable model determined by this procedure must possess both a reasonable glycosidic bond angle and a sufficiently low conformational energy.

2.2. Stereochemical Model Refinement

The stereochemical refinement generally follows immediately after a suitable initial model has been established. In this refinement, the chain model is placed in the most probable position in the crystal lattice and its packing energy is minimized. The variables in this refinement include both conformational and crystal packing variables. The conformational variables include all bond lengths, bond angles, and torsion angles. The packing variables include positions of the helix axes in the unit cell and the rotational and translational positions of the helices relative to their axes. The total energy E is conveniently calculated using the expression:

$$E = \sum_i \sigma_i^{-2} (a_i - a_{io})^2 + W \sum_{ij} w_{ij} E_{ij} \tag{1}$$

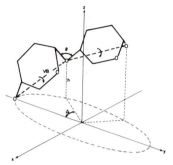

FIGURE 4   Residue rotation about
            the virtual bond (VB).

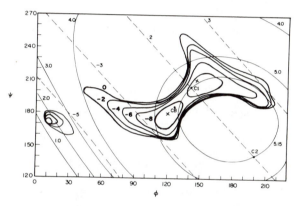

FIGURE 5   Conformational energy plot for cellulose (energy
            contours in kcal/mol, $\underline{n}$ - $\underline{h}$ contours are super-
            imposed, CB - cellobiose, C1 - cellulose I,
            x - global minima).

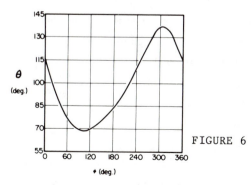

FIGURE 6   Variation of the bond
            angle θ with residue
            rotation about VB.

(where $\underline{a}_i$ is the value of a bonded parameter such as bond length, bond angle, torsion angle; $\underline{a}_{io}$ and σ_i are the corresponding standard value and its standard deviation, $\underline{E}_{ij}$ is the pairwise nonbonded energy describing the interaction between atoms $\underline{i}$ and $\underline{j}$, and $\underline{W}$ and $\underline{w}_{ij}$ are appropriate weights).

The total nonbonded energy $\underline{E}_{ij}$ may be calculated in various ways, e.g. from:

$$E_{ij} = E_{nb} + E_{hb} + E_{el} \tag{2}$$

(where $\underline{nb}$, $\underline{hb}$, and $\underline{el}$ refer to nonbonded, hydrogen-bonded, and electrostatic energies, respectively). Different expressions are available to calculate each component; however, in many cases an approximate $\underline{E}_{ij}$ can be conveniently and quickly calculated from the expression:

$$E_{ij} = \sigma_{ij}(d_{ij} - d_{ijo})^2 \quad \text{for } d_{ij} \leq d_{ijo} \tag{3}$$

(where $\underline{d}_{ij}$ is the nonbonded distance, $\underline{d}_{ijo}$ its most probable value, and σ_{ij} is an appropriate constant that can reflect such additional requirements as the hydrogen-bond or electrostatic energy). Although $\underline{E}_{ij}$ calculated from the above expression may be a pseudoenergy, it has been found to be a good approximation in most instances (4).

The result of such a stereochemical refinement is a lattice packing model that possesses reasonable stereochemical features. In some instances more than one reasonable model may be found, in which case all such models must be carried through the third and final refinement step.

2.3. Final Refinement

In the final step, the structure is solved through the refinement against the observed diffraction intensities (F_{obs}), while maintaining a stereochemical control over it.

The same variables are refined as in the stereochemical refinement step, but the quantity that is minimized now is usually the weighted crystallographic residual, $\underline{R}''$:

$$R'' = [\sum_i (w_i \, ||F_o| - |F_c||)^2 / \sum_i w_i |F_o|^2]^{1/2} \qquad (4)$$

(where $\underline{F}_o$ and $\underline{F}_c$ are the observed and calculated structure factor amplitudes, respectively, and $\underline{w}_i$ are the weights of individual diffracted intensities). In order to maintain reasonable stereochemistry during the final phase of refinement, the total packing energy $\underline{E}$ (cf. eq. 1) is simultaneously minimized. This is usually accomplished by minimizing the quantity Φ, a linear combination of $\underline{R}''$ and $\underline{E}$:

$$\Phi = fR + (1 - f)E \qquad (5)$$

(where the fraction $\underline{f}$ is so chosen that the energy term is not predominating, but is sufficiently large to maintain correct stereochemistry).

The computational procedures used in the above refinement system have been assembled in the computer program "PS79". The procedural and computational details pertaining to its use have been previously described (4).

More than 20 different crystal structures of cellulose and amylose and their derivatives have been examined using the above and similar methods of analysis, in this laboratory as well as by others. The structures having a particular bearing on the properties, functions, or biosynthesis of cellulose and starch are briefly reviewed in the following.

3. RESULTS

3.1. Cellulose

It has been known for a long time that cellulose can adopt four polymorphic crystalline forms -- celluloses I, II, III, and IV. Only cellulose I is the native form, while all oth-

ers are "man-made". Of the latter, cellulose II is the most
important as it is the form of mercerized and regenerated
celluloses. The interconversions between the polymorphic
forms have been studied extensively, in particular the cellulose I to II conversion as it occurs during the process of
alkali mercerization. These studies have revealed that the
native cellulose I form is metastable with respect to cellulose II and that the conversion from I to II is not reversible. It has become apparent that the crystal structure of
cellulose I results only through biosynthesis.

In order to clarify the structural aspects of these
observations, the crystal structures of celluloses I and II
(as well as III_I, IV_I, and IV_{II}) have been determined (5-8).
The results show that the polymorphic crystal structures of
cellulose fall into two families which differ in chain
polarity -- the parallel-chain family containing cellulose I
and some of its conversion products, and the antiparallel-
chain family containing cellulose II and some of its conversion products (cf. Fig. 7). The conversions within each
family are reversible, while the interconversions between
families are possible only in the direction from parallel to
antiparallel. From the point of view of stability, the
crystal structures of celluloses I and II (cf. Fig. 8)
clearly show one of the reasons for this irreversibility of
conversion. As can be seen in Fig. 8, the conformations of
the cellulose chains are essentially identical in both forms
-- extended and marked by intramolecular hydrogen bonds.
Likewise, in both structures the chains aggregate into similar parallel-chain sheets stabilized by inter-chain hydrogen
bonds. The main difference between the two forms arises
from the threedimensional packing of the sheets -- an all-

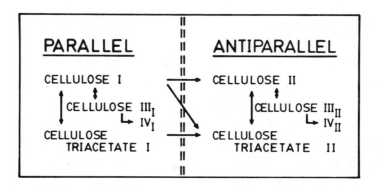

FIGURE 7   Parallel and antiparallel families of cellulose
            crystal structures.

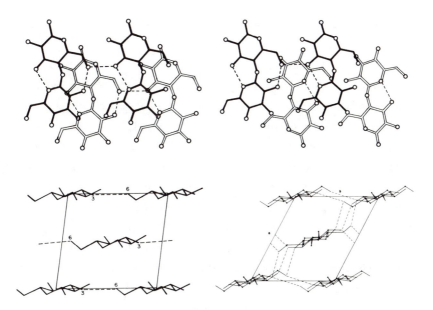

FIGURE 8   Top: Parallel (left) and antiparallel (right)
            sheet structures of celluloses I and II.
            Bottom: Unit cells of cellulose I (left) and
            cellulose II (right), projected down the
            fiber axis.

parallel arrangement of sheets in cellulose I, but an alternating, up-down packing of sheets in cellulose II. As is apparent in the a - b projections in Fig. 8, the parallel packing of cellulose I shows no hydrogen-bonding between the sheets, whereas in cellulose II there is much more extensive hydrogen-bonding in three dimensions. This lends a considerably higher degree of stability to cellulose II in comparison with cellulose I. Therefore, aside from any other argument, statistical or otherwise, concerning the cellulose I to II conversion, it is not likely that the cellulose II structure would spontaneously convert to cellulose I. By the same token, once the cellulose I crystal structure is disrupted, it will always recrystallize as the more stable cellulose II.

In this light, it is now much clearer how the cellulose I structure forms during biosynthesis. As shown by Brown and coworkers (9), the bacterium Acetobacter xylinum synthesizes cellulose in the form of multiple-chain microfibrils, "extruded" by the bacterium (cf. Fig. 9). The microfibrils crystallize before drying, and probably soon after synthesis. If a microfibril forms from a number of chains that are being simultaneously extruded in the same direction as parallel chains, which now appears to be very likely, the chains crystallize in a structure that is most easily accessible to them, i.e., the parallel-chain structure. The process is thus similar to one in which kinetic control rather than thermodynamic control is dominating.

A similar argument can be extended to the polymorph interconversions within each family. For example, when cellulose I is treated with liquid ammonia under nonswelling conditions, the parallel-chain cellulose III$_\text{I}$ structure results (7). Simply heating the latter in water (also a nonswelling condition) restores the structure of cellulose I. Apparently, ammonia is able to penetrate the fiber (most likely through amorphous regions separating the crystalline microfibrils) and convert the parallel-chain crystal without decrystallizing it first. The result is another parallel-chain structure.

Essentially the same mechanism may be responsible for some derivatives of cellulose. For example, cellulose I can be converted to cellulose triacetate (CTA) in two different ways. Under nonswelling conditions a parallel-chain CTA I is obtained, while under swelling conditions an antiparallel CTA II results. The crystal structures of both have been determined (10,11). It is interesting that when CTA I is swollen (such as by high temperature steam treatment) it converts irreversibly to CTA II, thus behaving in an analogous fashion with underivatized celluloses.

The observed conversions, particularly those occurring within a given chain polarity family, suggest a fiber morphology for cellulose in which amorphous or less crystalline regions are present. This is consistent with an incomplete degree of crystallinity characterizing most celluloses. In order to obtain a more detailed picture of the nature of such a morphology, a crystallographic study of ramie cellulose -- in the process of being mercerized by alkali -- was undertaken. It has been known for some time that a series of crystalline alkali-cellulose complexes can be obtained during the treatment of cellulose with alkali. The objective of the study of these complexes was to obtain information both on the mechanism of mercerization and the fiber morphology.

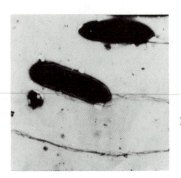

FIGURE 9   Cellulose microfibrils
            synthesized by Aceto-
            bacter xylinum. (Photo
            courtesy of Dr. R.M.
            Brown).

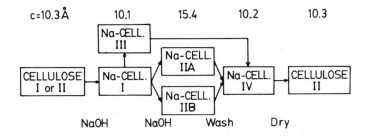

FIGURE 10   Na-cellulose crystal structures observed during
             mercerization of cellulose.

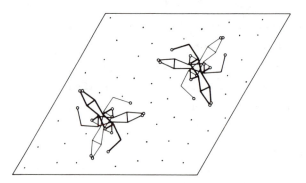

FIGURE 11   Probable structure of Na-callulose IIB, in pro-
             jection down the chain axis.

By slowing down the process of mercerization it was
shown that five intermediate crystal structures -- named
Na-cellulose I, IIA, IIB, III, and IV -- could be reproduci-
bly obtained during the conversion of cellulose I to cellulose II by the action of NaOH (12,13). The conversion pathways are shown in Fig. 10. It was also shown that during a
given conversion step, e.g., from cellulose I to Na-cellulose I, the transformation was a solid state one -- from one
crystal phase to another and not one in which the formation
of the ensuing crystal structure was preceded by the decrys-
tallization of the original structure. Mixtures of two
crystal structures could be readily observed at intermediate
times during the conversion. Particularly interesting was
the first conversion step (cellulose I to Na-cellulose I),
because when it was arrested at approximately 50% conversion
and the alkali was quickly washed out of the fibers, the
result was an approximately equal mixture of crystal structures of celluloses I and II. This result clearly suggests
that an antiparallel structure already forms during the
first step of mercerization. After the completion of the
first step, more alkali enters the crystal structure and the
latter expands, containing up to 60% NaOH and water in the
unit cell of Na-cellulose IIB. The chain conformation of
the cellulose in the latter structure is a three-fold helix
and not the familiar, ribbon-like, two-fold helix common to
all cellulose polymorphs. There are no intra- or interchain
hydrogen bonds present in this structure, shown in projection in Fig. 11. It is probable that any alkali-mediated
cellulose derivatization would proceed more rapidly when the
cellulose is in this stage. Washing out the alkali produces
a contraction of the lattice and a subsequent formation of

the cellulose II structure. The solid state conversion process is schematically illustrated in Fig. 12.

Monitoring crystallite sizes of both cellulose I and Na-cellulose I during the first conversion step produced the result shown in Fig. 13 (14). The Na-cellulose I crystal phase begins to form early, prior to any detectable decrease in the crystallite size of cellulose I. Only after the conversion is approximately 50% completed will the crystallites of cellulose I begin to diminish in size and the Na-cellulose I crystallites approach their final size. These results suggest the presence of amorphous regions of approximately 30–40 Å in lateral dimension, interspersed between approximately 60 Å wide crystallites (microfibrils) of cellulose I. Alkali should readily penetrate these amorphous regions converting them to crystalline Na-cellulose I. When the latter have grown to a sufficiently large size they will begin to peel chains from the adjacent cellulose I crystallites (cf. Fig. 12). Because the crystal structure of Na-cellulose I is already based on antiparallel chains, this process suggests a fiber morphology illustrated in Fig. 14. In this structure, a fiber contains a large number of crystalline microfibrils in which the chain directions are parallel. But because a fiber is most likely only a simple aggregate of microfibrils and is not formed as a result of a directed process such as is responsible for the formation of a microfibril, there should be, on the average, an equal number of "up" and "down" pointing microfibrils randomly aggregated into the fiber. The interface region between such an assembly of microfibrils may constitute a major fraction of the amorphous region. Because the aggregation of microfibrils is random with respect to polarity,

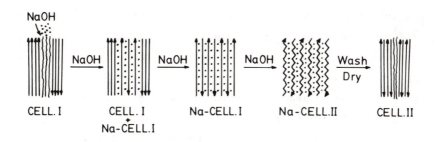

FIGURE 12   Probable transformation mechanism of crystal
             structures during cellulose I mercerization.

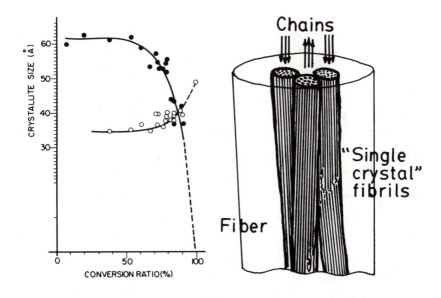

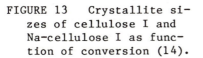

FIGURE 13   Crystallite sizes of cellulose I and
Na-cellulose I as func-
tion of conversion (14).

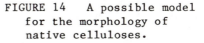

FIGURE 14   A possible model
for the morphology of
native celluloses.

the interfacial amorphous region will contain chains of both
"up" and "down" polarity. The creation of an antiparallel
Na-cellulose I structure in these regions is thus a natural
process. Only segments of adjacent chains need to move
short distances laterally, akin to a process of interdigita-
tion, in order to reach crystalline registry. The necessity
for a topologically cumbersome process of "turning chains
upside down" in order to form the antiparallel structure, is
therefore obviated.

3.2. Amylose and Starch

Contrary to cellulose whose fibrous morphology and excellent
crystalline orientation make it a nearly ideal substrate for
x-ray diffraction studies, starch is a very difficult
material to study by this method. Although most starches
are naturally crystalline, their degree of crystallinity is
considerably lower than that of celluloses. This, coupled
with spherulitic granular morphology, results in only powder
x-ray diagrams of relatively poor resolution. Moreover,
because most starches are molecularly composite polysaccharides -- consisting of amylose, a minor linear fraction, and
amylopectin, the major branched fraction -- recrystalliza-
tion of whole starch in a form that is more suitable for
x-ray diffraction is not possible.

Fortunately, it was discovered long ago that the linear
amylose can effectively be used as a model substance for
starch crystallinity. It can be fractionated into a relatively pure state, drawn into fibers (although not by a simple procedure in all cases), and studied in the latter form
by diffraction methods. The crystallinity exhibited by amylose can duplicate all of the native crystalline starch

forms, as well as show additional forms. In fact, as shown below, the crystalline polymorphism of amylose exceeds even that of cellulose. For all of these reasons, almost every-thing that has been learned about the crystallinity in starch has been through studies on amylose.

The crystal structures of amylose can generally be divided into two classes: (1) the single-stranded helical V-amyloses, and (2) the double-stranded (duplex) A-, B-, and C-amyloses (1). Both types of structures have a bearing on the morphology of native granular starches.

3.2.1. V-amyloses. When amylose is dissolved in a suitable solvent such as DMSO, the solution can be dried in the form of a film and the latter can be easily stretched into a fiber. The resulting crystal structure (shown in projection in Fig. 15) is marked by a compact helical conformation which exhibits a sizable channel in the middle of the helix. The channel is occupied by molecules of both DMSO and water (15). The structure is typical of many V-amylose complexes in which the complexing molecules can be alcohols, ketones, iodine, and many other small "guest" molecules. Generally, these complexing substances can be displaced by water, form-ing a series of V-hydrate amyloses. The lowest hydrate (confusingly named "V-anhydrous amylose") contains two water molecules for every three glucose residues, with the majori-ty of the water residing in the helix channel (cf. Fig. 15) (16). Regardless of the type of complexing molecule, the chain conformation -- shown in Fig. 16 -- is in the form of a rather wide helix, with typical n values of 6-8 residues/turn and h of ~1.3 Å. All of the V-amylose complexes ana-lyzed to date possess left-handed helical conformations.

The V-iodine amylose structure (17) is responsible for the deep blue color exhibited by starch granules or solutions when they are stained with iodine.

It has also been known for some time that when aqueous amylose solutions are saturated with a suitable complexing substance (such as n- or tert-butanol), rather well-formed V-complex amylose crystals are obtained. When this procedure is used with a well-fractionated, low degree of polymerization amylose (D.P. 50-100), the single crystals that are obtained yield excellent electron diffraction diagrams (18). The latter should provide high quality data that could lead to a detailed structure refinement.

When amylose is complexed with an ionic substance, such as alkali or a salt, a somewhat different single-stranded helical structure is obtained. A typical case is the KOH-amylose complex which is obtained when amylose triacetate is deacetylated in an alcoholic solution of KOH (19). Its structure (cf. Fig. 17) is based on a more stretched-out helix (with n = 6 and h = 3.735 Å) in which there is no channel in the center of the helix.

All of the single-helical structures are marked by a relative absence of interhelical hydrogen bonds, particularly in the case of V-amyloses. Undoubtedly, this contributes to their ease of solubilization in water.

3.2.2. A-, B-, and C-Amyloses.   All granular starches exhibit one of three distinct x-ray diffraction patterns -- the A, B, or C pattern. The A pattern is generally associated with cereal starches, the B pattern with tuber starches, and the C pattern is rare. It is not possible to obtain fibers of amylose corresponding to these three forms start-

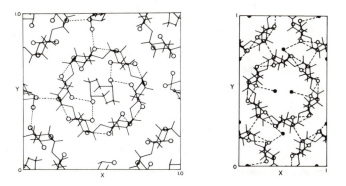

FIGURE 15   The crystal structures of "V–DMSO" (left) and
             "V–anhydrous" (right) amyloses, in projection
             down the helix axes. (Black dots signify water
             molecules).

FIGURE 16   The chain conformation of
             V–amyloses.

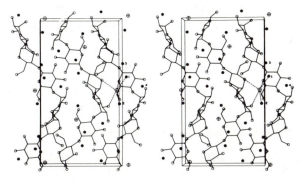

FIGURE 17   Stereo views of the crystal structure of the
             KOH–amylose complex (19).

ing with V-amyloses. However, the KOH-amylose complex, when suitably treated with high relative humidity followed by heating in water, will convert -- in the solid state -- to A-, B-, or C-forms, depending on the treatment. The resulting x-ray fiber patterns are sufficiently well resolved to have allowed a detailed refinement of the structures (20-22). The patterns are also identical in content with the respective powder patterns recorded from A-, B-, and C-starches.

The structures of A- and B-amyloses are illustrated in Fig. 18. The helix conformations are essentially identical in both and are based on a duplex (double-stranded) structure in which both chains are in a parallel orientation. The duplex is stabilized by numerous inter-chain hydrogen bonds. The main difference between the two structures resides in the packing of the duplexes -- a hexagonal packing with a "hole" in the center of the hexagon in B-amylose, and a similar packing with the hole filled by a duplex in A-amylose (cf. Fig. 18). In both cases, interstitial spaces between the duplexes are filled with water -- particularly the "hole" in the B-amylose structure. The amounts of water in the two crystal structures are in agreement with the results of hydration measurements. The C-structure is simply a mixture of the A and B crystal structures.

The duplex structure accounts well for all of the properties of starch and amylose. Both are insoluble in water and do not solubilize completely unless heated to at least 160° C. The extensive degree of hydrogen-bonding in the duplex must be a contributing structural feature to this property. When a solution of starch or amylose -- in at least a few percent concentration -- is allowed to cool, it

will gel. The higher the concentration of amylose, the
stronger will be the gel. After some time, the gel becomes
harder, will exhibit syneresis, and will eventually crystallize. This property is well known to starch and food chemists as "retrogradation". The crystalline pattern exhibited by a retrograded gel is the B-pattern. The gel cannot be
completely solubilized unless heated again to 160° C. These
observations suggest the formation of duplex structures as
the solution gels, and a crystalline packing of the duplexes
in the retrograded state. This course of events may occur
as illustrated in Fig. 19.

The structure of the starch granule is more complicated, but again, the duplex crystal structure is in good
agreement with its characteristics. For example, all starch
granules show approximately the same degree of crystallinity
irrespective of their amylose content (even in waxy varieties which do not contain amylose). When amylose-
containing starch granules are leached with hot water, the
amylose component of the granule is selectively solubilized.
When the granules of amylose-containing starch varieties are
stained with iodine, a blue color develops (waxy varieties
do not stain blue). All three properties suggest that the
amylopectin fraction is largely responsible for crystallinity, most likely due to duplex formation in its longer
branches. The amylose fraction is probably mostly amorphous
or in a disordered, V-type, single-helical conformation.
Some fraction of it may be co-crystallized with amylopectin.
The amorphous or V-type conformation can be easily solubilized by water leaching and may also be responsible for the
blue iodine color. The relatively porous duplex crystal
structure -- particularly the B-form with its channel within

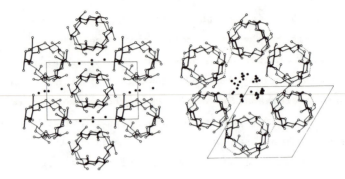

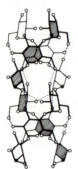

FIGURE 18   Projections of the crystal
            structures of A- and B-amyloses (top
            left and right).   Conformation of the
            duplex helix (left).   (Black dots sig-
            nify water molecules) (20-22).

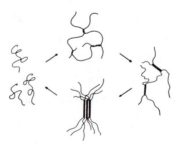

FIGURE 19   Probable mechanism of amylose gelation and
            retrogradation.

the hexagonal packing -- is also in agreement with the ease
with which starch granules can be hydrated, dehydrated, or
penetrated by various liquids.

The one structural characteristic of amylopectin -- why
it is crystalline in the native granule but cannot be crystallized once it has been solubilized -- has not been clarified by these diffraction studies. Obviously, biosynthetic
mechanisms may have a role here and they must be studied by
other means. However, it is not unlikely that some well-
designed diffraction experiments may lend help to these
studies also.

Acknowledgment. These studies have been supported by
the National Science Foundation under grants No. CHE7501560,
CHE7727749, and CHE8107534, and by the NATO grant RG1386.

REFERENCES

1. A. SARKO and P. ZUGENMAIER, in Fiber Diffraction
    Methods, ACS Symp. Ser., 141, 459 (1980).
2. A. SARKO, Tappi, 61, 59 (1978).
3. D.A. BRANT, Ann. Rev. Biophys. Bioeng., 1, 369 (1972).
4. P. ZUGENMAIER and A. SARKO, in Fiber Diffraction
    Methods, ACS Symp. Ser., 141, 225 (1980).
5. C. WOODCOCK and A. SARKO, Macromolecules, 13, 1183
    (1980).
6. A.J. STIPANOVIC and A. SARKO, Macromolecules, 9, 851
    (1976).
7. A. SARKO, J. SOUTHWICK, J. HAYASHI, Macromolecules,
    9, 857 (1976).
8. E.S. GARDINER and A. SARKO, Can. J. Chem. (in press).
9. C.H. HAIGLER, R. M. BROWN, Jr., M. BENZIMAN, Science,
    210, 903 (1980).

10. A.J. STIPANOVIC and A. SARKO, Polymer, 19, 3 (1978).
11. E. ROCHE, H. CHANZY, M. BOUDEULLE, R.H. MARCHESSAULT,
     P. SUNDARARAJAN, Macromolecules, 11, 86 (1978).
12. T. OKANO and A. SARKO, J. Appl. Polym. Sci, (in
     press).
13. T. OKANO and A. SARKO, J. Appl. Polym. Sci., (in
     press).
14. H. NISHIMURA and A. SARKO (to be published).
15. W.T. WINTER and A. SARKO, Biopolymers, 13, 1461 (1974).
16. W.T. WINTER and A. SARKO, Biopolymers, 13, 1447 (1974).
17. T.L. BLUHM and P. ZUGENMAIER, Carbohydr. Res., 89, 1
     (1981).
18. F.P. BOOY, H. CHANZY, A SARKO, Biopolymers, 18, 2261
     (1979).
19. A. SARKO and A. BILOSKI, Carbohydr. Res., 79, 11
     (1980).
20. H-C.H. WU and A. SARKO, Carbohydr. Res., 61, 7 (1978).
21. H-C.H. WU and A. SARKO, Carbohydr. Res., 61, 27 (1978).
22. A. SARKO and H-C.H. WU, Starch, 30, 73 (1978).

MOLECULAR INTERACTIONS OF POLYSACCHARIDES AND THEIR

RELATIONSHIP TO BULK PROPERTIES

W.T. WINTER
Department of Chemistry
Polytechnic Institute of New York
Brooklyn, New York 11201 USA

Polysaccharide associations may be homologous or heterologous. The former often lead to complex morphologies such as the cellulose microfibril and the multistranded helices of amylose, (1→3) glucans, hyaluronate, etc. Interactions within and between these domains determine mechanical, rheological and solubility properties of the polymers. The heterologous interactions involve polysaccharide associations with ions, solvent molecules, or other polymeric species. In this paper examples of each type of interaction will be considered along with some of the methods available for their study.

1. HOMOLOGOUS INTERACTIONS IN POLYSACCHARIDES

Although many spectroscopic methods have been employed in studying the solid-state associations of polysaccharides, X-ray diffraction and electron microscopy continue to provide the richest source of data. The structures which have been observed can be broadly classified as either multistranded helices or lateral associations. The first class includes the native A- and B-amyloses[1], hyaluronate[2], and ι-carrageenan[3], all of which form double helices, as well as the β(1→3) glucans such as curdlan which forms a 3-fold helix[4]. By far the best example of a structure arising from lateral associations of poly-saccharides chains is the microfibril of cellulose and other $\beta(1\rightarrow4)$ glycans. In both instances the dominant attractive interactions which stabilize the structure are thought to be hydrogen bonds. The question of which hydrogen bonds will be selected from the myriad choices possible in a given polysaccharide is not yet understood at a level that would permit prediction of chain associa-tion modes solely from a consideration of primary struc-ture. However, the existing experimental data clearly point to the primacy of main chain conformation over intermolecular association. Thus, cellulose, cellulose triacetate, mannan, and chitin, all of which are diequa-torially linked $\beta(1\rightarrow4)$ polymers, have the same funda-mental conformation, a 2-fold helix with a pitch of about 10.4 Å, although they pack into a wide variety of crys-talline domains. In fact this general conformation is preserved even in the more extreme cases of gluco-mannans[5], where glucose and mannose substitute for one another in an apparently random fashion along the linear chain, and galactomannans such as tara, guar and carob gums[6] which have varying amounts of $\alpha(1\rightarrow6)$ galactose substituents attached to the mannan main chain.

What all of these findings suggest to this author is that although presently available techniques offer at least some chance for predicting polysaccharide conforma-tion, there is almost no chance of predicting its packing or self-association. Since the compounds are available it is, of course, possible to determine both conformation and self-association modes when such knowledge is deemed necessary.

From an academic's perception of the industrial world, this inability is not a serious flaw. Almost any conceivable industrial process involving polysaccharides will require mixing them with other molecular species, e.g. ions, water, other solvents, or even other macromolecules. As a result, a more important question than that of predicting measurable structures is to predict the perturbations arising from interactions of existing structures with other molecular species. For polysaccharide systems currently in commercial use such perturbations can have origins as diverse as the swelling of cellulose by alkali in the manufacture of paper and the immunogenic reactions of capsular polysaccharides from bacteria with human antibodies, which facilitates the use of polysaccharide vaccines.

2.   HETEROLOGOUS INTERACTIONS INVOLVING POLYSACCHARIDES

2.1 Polysaccharide - Solvent Interactions

In characterizing polysaccharide solids containing solvent molecules, there are at least three basic questions to answer. How much solvent is present within the polymer matrix? To what extent is the solvent species bound to the matrix? And finally, how does the presence of solvent molecules affect the organization of the matrix?

The amount of solvent present in the system can often be determined with sufficient accuracy from either classical gravimetric methods or from thermogravimetric analysis (TGA). It is important, however, to verify that the loss of mass is really a consequence of solvent

volatility. This may be accomplished either by showing
that the process is reversible or by analytical procedures designed to demonstrate the chemical integrity of
the polymer.

An alternative approach which provides a lower limit
to the solvent content of crystalline domains involves
measuring the density, ρ, of the bulk material and the
lattice constants of the unit cell. Together with
Avogadro's number, these data allow us to calculate a
gram molecular mass for the unit cell from the equation

$$\rho = M/(N_{av} V) \tag{1}$$

where V is the unit cell volume derived from the lattice
constants. This gram molecular mass must then be rationalized in terms of the mass of the polysaccharide repeating motif and the mass of the solvent molecule.

As part of an ongoing study of the capsular polysaccharide from Xanthomonas campestris, xanthan gum, we
have measured the water content for films of the polysaccharide stored under various constant humidity conditions using X-ray methods as well as TGA and differential scanning calorimetry, DSC. In this latter approach,
the assumption is made that the endothermic peak arises
from the vaporization of water in the films. As shown in
Table I the agreement among these three indices of water
content is surprisingly good.

Thermal analysis also provides a simple approach to
the problem of ordered versus mobile solvent. Here one
makes the initial assumption that mobile or unbound water
should have essentially the same physical properties as
bulk water[7]. Thus, one expects an exothermic transition
similar to the latent heat of fusion for the bulk solvent

TABLE I.

Water content of xanthan gun by X-ray, DSC and TGA methods

	X-RAY*	DSC	TGA
% water (w/w)	24	32	35
water mol/pyranose	3.2	5.0	5.7
initial r.h. (%)	77	92	92

* The X-ray result is based upon an orthorhombic
  unit cell with a = 31.3 Å, b = 17.8 Å, c = 48.8 Å
  ρ = 1.514 g/cm^3.

and quantitation of the area under this curve provides a
measure of the fraction of mobile solvent.

Alternatively, it may be possible to differentiate
site bound from mobile solvent molecules by their different effects on the intensity distribution in an x-ray
diffraction pattern. For example, site bound water[8] and
DMSO[9] molecules have been located in polysaccharide crystal structures by the method of Fourier difference
synthesis.

The effect of mobile solvent on diffraction data can
also be predicted through the use of solvent-weighted
atomic scattering factors[1,10]. Their use replaces the
conventional scattering factor calculated for an N electron point atom in vacuo by an N electron atom occupying
a finite volume in a fluid whose uniform electron density
corresponds to that of the bulk solvent. While this is a
highly subjective procedure, it is most certainly prefer-
able to the atom in a vacuum model when considering polysaccharide structures which may contain upwards of 40%

solvent on a weight basis.

Using the solvent-weighted scattering factor approach it has been observed that polysaccharides frequently organize on an open hexagonal net giving rise to a honey-comb-like array with large columnar channels filled with solvent molecules. Examples of this kind of structure in Na chondroitin sulfate and B-amylose are shown in Fig. 1.

Two additional techniques which have yet to be use-fully applied to polysaccharides but which may prove fruitful are ^{2}H NMR and neutron scattering. The former can be used to examine both the orientation and mobility of deuterated solvents through analysis of the line pro-files and intensities of the ^{2}H solid echo spectra[11]. The neutron scattering approach has already had dramatic impact on the synthetic polymer field, since it is the only technique which permits an evaluation of molecular dimensions in non-crystalline solids. For polysaccha-rides, their interactions with surfactants and in nematic liquid crystalline phases with molecules like N-methyl morpholine oxide are two areas where the neutron techni-ques may provide critical insight[12].

2.2 Polysaccharide - Ion Interactions

The X-ray methods previously discussed in terms of poly-saccharide-solvent interactions have been used in defin-ing the structures of ionic complexes. Most of our knowledge to date deals with one of three polysaccharide classes; the glycosaminoglycans, or GAG's[2,8], amylose[13], or cellulose[14]. For anionic polysaccharides, the X-ray studies on GAG's reveal specific ion-binding sites which may, in some instances, arise from conformational transi-

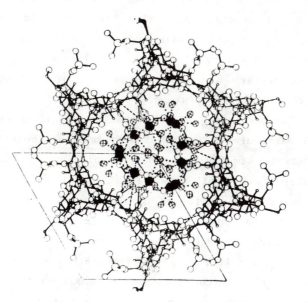

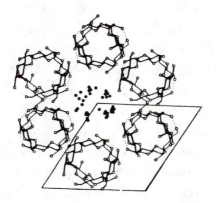

Figure 1 Water channels formed by hexagonal packing in a)
Na chondroitin sulfate (from ref. 10) and b) B-amylose
(from ref. 1)

tions in response to ion exchange. Mechanical properties such as gel strength have also been found to vary for certain anionic polysaccharides as a funtion of cationic species[15]. However, a precise correlation of the relationship between ion-binding and mechanical properties has yet to be undertaken.

The mercerization of cellulose in alkali is evolving into a particularly fascinating example of the relationship between structure and processability. Although the molecular shapes of the initial and final compounds (cellulose I and cellulose II) are quite similar, their crystalline packing is totally different, involving an array of parallel chains in the former and anti-parallel chains in the latter. The intermediate alkali cellulose structures show a conformational transition to a 3-fold helical form not unlike those observed from nematic mesophases of hydroxypropyl cellulose or cellulose tricarbanilate. Based on their X-ray data, Sarko and his co-workers have developed a model for the mercerization process involving the progressive solvation of chains from the surface of cellulose I crystallites, forming first the alkali cellulose intermediates and ultimately new crystallites of cellulose II. Since the cellulose I crystallites need not be parallel this provides a rational explanation for the conversion from a parallel chain crystallite to an anti-parallel chain crystallite without a concomitant loss of cellular structure[14].

2.3 Polysaccharide Associations with Other Polymers

The last and potentially the most rewarding area of investigation relates to the interaction of a polysaccharide molecule with another polymer - polysaccharide or otherwise. Such interactions may be hightly specific as with those involving enzyme-substrate or antibody-antigen interactions, or comparatively non-specific and occurring only on the surface of crystallites.

In our laboratory at Brooklyn and previously in collaboration with S. Arnott at Purdue we have attempted to form crystalline complexes of galactomannans such as guar, tara and carob gums with xanthan gum. Although the existence of such complexes in solution has been widely reported[15] and we were able to duplicate the solution experiments, X-ray data obtained from crystalline films of the 'complexes' invariably were indistinguishable from those of native xanthan. Similar experiments with the κ-carrageenan - gal-carob gum system were performed by V.J. Morris and co-workers and again the X-ray patterns were characteristic of a single phase, κ-carrageenan[16]. They were, however, able to show that fracture stress in the gels did follow the same pattern as solution viscosity with a maximum at a 2:1 ratio of carrageenan to carob.

One final mixed polysaccharide system is that of alginate, preferably a guluronate-rich fraction, with pectinic acid. Like the galactomannan systems previously mentioned, this system has been shown to gel under conditions (pH<3.1, 0.2% total polysaccharide and 1:1 ratio of polymers) where neither component would gel alone. We have recently obtained an X-ray pattern from such a complex which has the same layer line spacing as the pure alginate but a significant difference in the intensity distribution along the equatorial layer line. These data are strongly suggestive that we have obtained the complex

in the solid state and further experiments are in progress to verify this result and confirm the nature of
the complex.

3. CONCLUSIONS

When I agreed to make this presentation, I had rather
grandiose views of what could be done in a short presentation. Instead I now hope that I have at least
managed to highlight some areas of current interest and
new techniques for understanding the myriad ways in
which polysaccharides can associate with themselves and
other molecular species.

REFERENCES

1. H.C. WU and A. SARKO, Carbohydr. Res., 61, 7-25 and
   27-40 (1978).
2. J.K. SHEEHAN, K.H. GARDNER and E.D.T. ATKINS,
   J. Mol. Biol., 95, 113-135 (1977).
3. S. ARNOTT, W.E. SCOTT, D.A. REES and C.G.A. MCNAB,
   J. Mol. Biol., 90, 253-267 (1974).
4. R.H. MARCHESSAULT, Y. DESLANDES, K. OGAWA and P.R.
   SUNDARARAJAN, Can. J. Chem., 55, 300-303 (1977).
5. H.D. CHANZY, A. GROSENRAUD, J.P. JOSELEAU, M. DUBE
   and R.H. MARCHESSAULT, Biopolymers, 21, 301-319
   (1982).
6. a) R.H. MARCHESSAULT, A. BULEON, Y. DESLANDES and T.
      GOTO, J. Colloid Interface Sci., 31, 375-382
      (1979).
   b) W.T. WINTER, Y.Y. CHIEN and H. BOUCKRIS, Struc-
      tural Aspects of Food Galactomannans, Gums and
      Stabilisers for the Food Industry 2, Editors G.O.
      Phillips, D.J. Wedlock and P.A. Williams (Perga-
      mon Press, Oxford, 1984).
7. H.S. FRANK and W.Y. WEN, Disc. Faraday Soc., 23,
   133-140 (1957).
8. J.M. GUSS, D.W.L. HUKINS, P.J.C. SMITH, W.T. WINTER,
   S. ARNOTT, R. MOORHOUSE and D.A. REES, J. Mol. Biol.,

95, 359-384 (1975).

9. W.T. WINTER and A. SARKO, Biopolymers, 13, 1461-1482 (1974).

10. W.T. WINTER, S. ARNOTT, D.H. ISAAC and E.D.T. ATKINS, J. Mol. Biol., 125, 1-19 (1978).

11. H.W. SPIESS, Colloid Polym. Sci., 261, 193-209 (1983).

12. W.F. BRINKMAN, Current Studies of Neutron Scattering Research and Facilities in the United States (National Academy of Science, Washington, D.C., 1984).

13. A. SARKO and A. BILOSKI, Carbohydr. Res., 79, 11-21 (1980).

14. A. SARKO, T. OKANO and H. NISHAMURA, Mercerization of Cellulose: What can Crystallography Tell Us?, 1983 International Dissolving and Specialty Pulps Conference (TAPPI, TAPPI Press, Atlanta, 1983).

15. E.R. MORRIS, Rheology of Hydrocolloids, Gums and Stabilisers for the Food Industry 2, Editors G.O. Phillips, D.J. Wedlock and P.A. Williams (Pergamon Press, Oxford, 1984).

16. V. CARROLL, M.J. MILES and V.J. MORRIS, Synergistic Interactions between Kappa Carrageenan and Locust Bean Gum, Gums and Stabilisers for the Food Industry 2, Editors G.O. Phillips, D.J. Wedlock and P.A. Williams (Pergamon Press, Oxford, 1984).

17. I.C.M. DEA, Alginate Pectin Interaction, Gums and Stabilisers for the Food Industry, Editors G.O. Phillips and P.A. Williams (Pergamon Press, Oxford, 1983).

ACKNOWLEDGMENT

The author wishes to acknowledge the support of the donors of the Petroleum Research Fund of the American Chemical Society.

PROCESSING POLYSACCHARIDES WITH MEMBRANES

J.R. VERCELLOTTI and S.V. VERCELLOTTI
V-Labs Incorporated
Covington, Louisiana 70433

ELIAS KLEIN
University of Louisville
School of Medicine
Louisville, Kentucky 40292

More recent methods of membrane characterization permit separation schemes to be devised with a high degree of reliability. Membranes so calibrated have been used in the present study to purify laminaran from seaweed; to prepare linear maltodextrins with degree of polymerization of glucose from three to ten in a sharp fraction with weight average molecular weight (M_w) 1500; and to fractionate a broad molecular weight (MW) starch hydrolysate into three products of 20,000-70,000 (11%), 2000-20,000 (27%), and 180-2800 (62%) MW. The success of such membrane separation processes must be considered in the light of the fluid dynamics of the system and equations are presented defining the calibration procedure.

1. INTRODUCTION

Recent Symposia[1-3], describing progress in membrane processes, attest to the penetration of membrane technology into industries where high energy costs or heat sensitive

chemicals present separation problems. Examples include
the preparation of foodstuffs and specialty chemicals in
the dairy, pharmaceutical, and metalworking industries.
The application of membranes to such industrial separations
has been facilitated by more simple membrane
characterization procedures and through a better
understanding of the dependence of attainable separations
upon the operating variables. Early industrial
applications centered on simple dehydration of protein
solutions, or on recovery of emulsified oils. The sieving
properties of available membranes were not well understood,
and only materials of highly divergent MWs were considered
suitable for separation by membranes. However, there are
now a number of methods for characterizing the response of
membranes to solution components[4-6], allowing separation
schemes to be devised with a high degree of reliability.

Two factors determine the feasibility of separating a
solute either from its solvent or from a mixture of solutes
in solution. The first is the intrinsic rejection (R) of
the membrane for a specific solute.

$$R = 1 - \frac{C_f}{C_m} \tag{1}$$

In equation (1) C_f is the solute concentration in the
filtrate, and C_m is the solute concentration on the
pressured side of the membrane. Equation (1) is applicable
to pressure driven separations operating without fluids in
the receiver, as would be the case for dialysis. Rejection
is a function of both membrane properties and the operating
variables, principally the transmembrane velocity. The
concentration of a specific solute in the filtrate (C_f) of

an ultrafiltration membrane (characterized by the two constants P_m and σ) is given by[7]:

$$C_f = C_m \quad 1 - \frac{\sigma(e^b - 1)}{(e^b - \sigma)} \qquad (2)$$

where $b = J_v(1-\sigma)/P_m$ (2a)

and $J_v = L_p(\Delta p - \sigma\Delta\pi)$ (2b)

In these equations J_v is the transmembrane velocity (or flux per unit area), Δp is the transmembrane pressure gradient, L_p is the hydraulic conductivity of the membrane, P_m is the diffusive permeability of the solute in the membrane, and σ is the Staverman reflection coefficient, a measure of the membrane's ability to exclude the solute. Inspection of equations (1) and (2) shows that R is a function of transmembrane velocity; as J_v increases the value of R asymptotically approaches the values of σ. The concentration, C_m, is the solute concentration at the interface between the membrane and the feed solution; it cannot be measured but must be estimated from analysis of the bulk composition (C_o) and mathematical approximations of the extent of concentration polarization adjacent to the membane surface. The value of $\Delta\pi$ is the theoretical osmotic pressure across the membrane.

The second factor governing separations by membranes is the nature of the concentration polarization layer. As a solution containing solutes is swept along a channel of height, 2t, at a velocity J_{BF}, with a transmembrane loss of solvent given by J_v, the solutes will be swept toward the membrane wall. If the solutes cannot permeate the membrane they will accumulate at the solution-membrane interface. The solute concentration at the interface will reach an equilibrium value determined by the solute's ability to diffuse back into the main channel, which is described by the diffusivity coefficient, D. The wall concentration, C_m, will now be greater than the bulk concentration, C_o, and this will affect the observed rejection, R_{obs}, which is based upon the ratio of the filtrate concentration to the measured bulk feed concentration. The observed rejection is related to the true rejection at the surface of the membrane (i.e., in the absence of concentration polarization) by[5]:

$$R_{true} = \frac{R_{obs} e^{KJ_v}}{(1 + R_{obs}(e^{KJ_v} - 1))} \tag{3}$$

Here, K, the boundary layer resistance, is calculated at a point Z along the flow channel by

$$K = 0.8456 (Zt/J_{BF}D^2)^{.333} \tag{4}$$

Equation (4) must be integrated along the length of the channel to yield the mean rejection for the device. Methods for these numerical procedures, or for simpler approximations, have been developed for flat channels[5] and for hollow fibers[8].

The effect of concentration polarization can be quite

dramatic.   Figure   1 shows the rejection as a function of transmembrane velocity for inulin using a polyacrylonitrile ultrafiltration membrane (20,000 MW nominal cutoff).

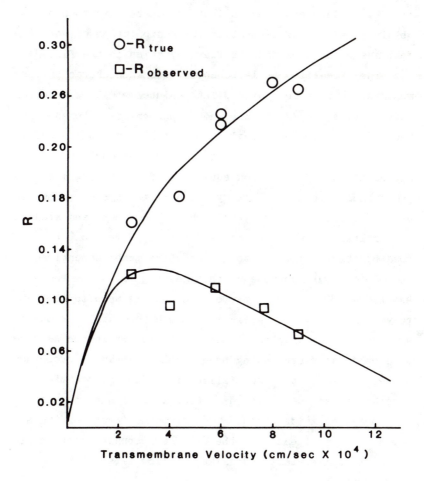

FIGURE 1   Rejection of inulin by polyacrylonitrile
membrane (20,000 MW nominal cutoff)

Two curves are shown; one for the observed rejection, which goes through a maximum, and one for the "true" rejection, corrected for the boundary layer effects. The true rejection asymptotically reaches a limiting value equal to σ. Such experiments indicate that it is possible to obtain a variety of separations by manipulation of both the membrane properties and the fluid dynamics in the filter.

The idealized relationship used when characterizing membrane properties with dilute solutes at low extraction ratios (i.e. J_v/J_{BF}) are not appropriate in predicting whether or not practical separations can be achieved. Rather, the process chemist or engineer must develop an appreciation of the consequences of combining membrane properties with the fluid dynamics of the system. The use of turbulence promoters, thin channel devices, and staging of filters provides an additional dimension to the possibilities available to the creative membrane engineer.

Since our purpose at this meeting is to see how such knowledge of membrane performance can be used to separate polysaccharides, we will illustrate some applications by showing three examples of polysaccharide isolation we have developed at V-Labs using predictable membrane performance data. Our laboratory specializes in custom preparations of carbohydrates; we have developed isolation processes for six polysaccharides based on membrane fractionation and would welcome ideas for others. The membranes that we use are all commercially available hollow fiber or plate and frame devices.

2. CARBOHYDRATE PREPARATION

2.1. Laminaran

The isolation of the soluble beta-D-(1 → 3)-glucan, laminaran, (MW 5000-7000) from seaweed is much simpler when good quality Laminaria digitata is available. However, given the structural problems involved with polysaccharides from mixtures of brown seaweeds, we attempted to use membranes to separate the soluble laminaran from these extracts. In Figure 2, we show our extraction and workup of the acid soluble portion of the seaweed extract. Much colloidal material is still present, which is removed with a cross-flow colloidal filter of 0.6 micron pore size. The ultrafiltrate contains all the soluble materials with MW less than ca. 1,000,000. At this stage, there are about 80 liters of extract which we simultaneously dialyze with a cellulose hollow fiber unit and concentrate with the reverse osmosis unit to about 40 liters. Dialysis lowers the salt level at the same time that the solution is being concentrated, thus lowering the osmotic backpressure of the solution. In order to separate all of the material from 50,000-80,000 MW from the laminaran, the extract is diafiltered again through a polyacrylonitrile membrane to recover the lower MW fraction in the filtrate. After dialysis with a nominal 1000 MW cutoff cellulose hollow fiber dialyzer, the filtrate is concentrated with the reverse osmosis unit and precipitated into ethanol. The final product, laminaran, has a M_W of 7500 and number average molecular weight (M_N) of 6800. By gas chromatography of the alditol acetates, the composition of this laminaran is 95% glucose with 3-5% branching according to ^{13}C nuclear magnetic resonance.

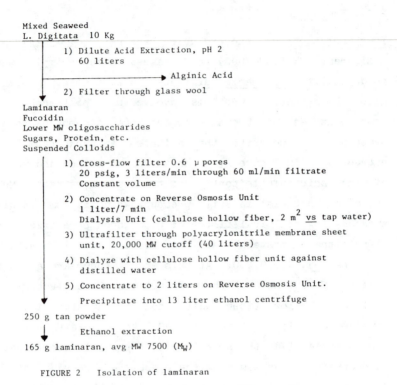

Mixed Seaweed
L. Digitata 10 Kg

   1) Dilute Acid Extraction, pH 2
      60 liters

          → Alginic Acid

   2) Filter through glass wool

Laminaran
Fucoidin
Lower MW oligosaccharides
Sugars, Protein, etc.
Suspended Colloids

   1) Cross-flow filter 0.6 μ pores
      20 psig, 3 liters/min through 60 ml/min filtrate
      Constant volume

   2) Concentrate on Reverse Osmosis Unit
      1 liter/7 min
      Dialysis Unit (cellulose hollow fiber, 2 m^2 vs tap water)

   3) Ultrafilter through polyacrylonitrile membrane sheet
      unit, 20,000 MW cutoff (40 liters)

   4) Dialyze with cellulose hollow fiber unit against
      distilled water

   5) Concentrate to 2 liters on Reverse Osmosis Unit.

      Precipitate into 13 liter ethanol centrifuge

250 g tan powder

      Ethanol extraction

165 g laminaran, avg MW 7500 (M_W)

FIGURE 2   Isolation of laminaran

2.2. Maltodextrin

A low MW starch hydrolysate product was required for testing industrial membranes according to Klein et al[4]. A Novo Industrie amylase-pullulinase system was used for hydrolysis. Figure 3 shows how membranes were put to work removing not only high MW enzyme, but also the yeast cells and debris from a glucose fermentation. The yeast cells are easily removed with the colloidal filter; residual enzyme is taken out with nominal 20,000 MW cutoff

polyacrylonitrile membrane, and the syrup is concentrated
by reverse osmosis and an osmotic salt concentrator
(V-Labs) to the final syrup, which contains 33% solids.
The maltodextrins are spread between maltotriose and
maltodecaose.

Starch (10 Kg)

α-Amylase (0.6 liters) (Novo)
  20 liters H_2O

Slurry and heat at 70°
  0.5 hours

Heat to 95°, 0.5 hours

Cool to 55°, Add 0.3 liters α-amylase
  0.45 Kg pullulanase (Novo)

Hold at $55-60^o$, 18 hours
  to dextrose reducing equivalent of 30

Adjust to 17% solids at 30^oC (60 liters) and Acidify,
  0.5 Kg yeast (S. cerevisiae) (Redstar)

pH 5.0 (maintain with NaOH) 30°, Air stream,
  mechanical stirrer 48 hours

Filter through 0.6 μ cross-flow colloidal
  filter at constant volume for 3 volumes

Ultrafilter through polyacrylonitrile membrane,
  20,000 MW cutoff

Concentrate to 40 liters with reverse osmosis
  unit

Concentrate to 25 liter with cellulose acetate
  osmotic concentrator (V-Labs)

DP 3-10 Avg MW ∿ 1500; 20-30% branching, yield 8 Kg.
  Syrup 33% solids (25 liters).

FIGURE 3   Linear maltodextrins, DP 3-10, from starch hydrolyzate;
             Novo process of hydrolysis

3. FRACTIONATION

Finally, we wish to discuss the fractionation of narrower
ranges of starch oligosaccharides from a broad range of
maltodextrins. This starting material is Maltrin M100 from
Grain Processing Corp., but equally useful for
fractionation are a number of other soluble starch
hydrolysates. This water soluble maltodextrin is 65% in
the 180-2000 MW range and 35% in the 2000-80,000 MW range.
Upon ultrafiltration with the nominal 20,000 MW cutoff
polyacrylonitrile membrane, the retentate is concentrated
to a product as in Figure 4.

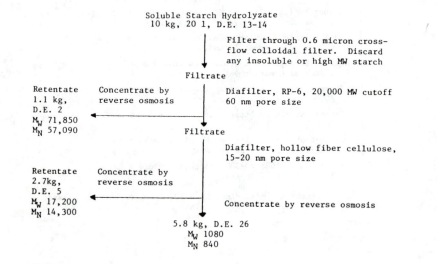

FIGURE 4   Isolation of maltodextrin fractions

This narrow fraction is completely water soluble and contains 20,000-70,000 MW material (11% by weight of starting material, dextrose equivalent (D.E.)= 2, M_W 71,850, M_N 57,090). The permeate is ultrafiltered through a 2000 MW-cutoff cellulose hollow fiber (ca 18nm pore size), to give two more fractions. The retentate has a broad 2000-20,000 MW (27%, D.E. 4, M_W 17,400, M_N 14,300), while the ultrafiltrate has a narrow MW range, 180-2800 MW (62% by weight, D.E. 26, M_W 1030, M_N 880).

In addition to the three processes described above, we have performed similar fractionations on commercial inulin (Pfanstiehl) with hollow fiber membrane units at 50°C to obtain three inulin fractions with M_W's of 6000 (30%), 4500 (50%), and 3500 (20%). Similarly, we have fractionated hemicellulose hydrolysates into products with various physical properties and MW ranges. Other separations have been performed with dilute solutions of carboxymethylcellulose, sulfated amylose, chitosan, N-carboxymethyl chitosan, xanthan gum, alginate, hydroxypropyl guar, pectin, and glycosaminoglycans.

Imaginative use of membranes can lead to new industrial products. The pharmaceutical industry has many applications of tangential flow membranes for isolation of antibiotics from fermentation broths. Scaling-up of the equipment does not present a problem either in cost or durability. The need for new products and their marketability will probably be the principal determinant to use of these processes.

REFERENCES

1. Symposium on Ultrafiltration Membranes and Applications, American Chemical Society, Washington, D.C., 1979.
2. Material Science of Synthetic Membranes, American Chemical Society, St. Louis, MO, 1984.
3. Membrane and Material Processes, European Society of Membrane Science and Technology and the Membrane Society of Japan, Stresa, Italy, June 18-22, 1984.
4. E. KLEIN, J. Memb. Sci. 15, 15-26, (1983).
5. R. P. WENDT, J. Memb. Sci. 5, 23-49, (1979).
6. A. MICHAELS, Separation Sci. Technol. 15, 1305-1322, (1980).
7. K.S. SPRIEGLER, Desalination 1, 311, (1966).
8. R.P. WENDT, Chem. Eng. Commun. 8, 251-262, (1981).

MODIFICATION OF THE PROPERTIES OF WATER

D.F. DURSO
Johnson & Johnson
East Windsor, New Jersey 08520

Since the beginning of time, man has used polysaccharides
to modify the properties of water in many useful ways,
usually for improving food and clothing. One of the largest
of the truly synthetic materials became universally employed
shortly after World War II when the synthetic detergent
industry erupted, in large measure because of the availability of CMC. During the past ten years, we have seen new
activity in the area of cellulose and starch polymers modified so that they have such useful properties that they
have become popularly known as "superabsorbents."

It has been known for many years that derivatives of
cellulose and starch can be prepared by grafting. In these
reactions, free radicals are generated within the substrate
by some suitable catalyst system. Usually the hydrogen of
the hydroxyl is eliminated and the oxygen atom serves as
the site of the free radical. In the presence of suitable
vinyl-type monomers, these active species will then cause
a closely associated synthetic chain molecule to be produced.
Whether the new molecule is attached covalently or by weaker
bonds, the result is a major modification of the water-

solubility of the natural substrate. The components of the
final product can be further modified by chemical reactions
generally known as "crosslinking." These can be as simple
as internal ester formation (between hydroxyl and carboxyl
groups), or they can be as complicated as those obtained
by chemical additions to double bonds. Very recently we
have seen the introduction of fully synthetic molecules
which also fall under the general heading of hydrogels.

These 1984 comments center on the lack of knowledge
concerning the mechanism by which these compounds are able
to modify the properties of water, especially viscosity.

For economic reasons, chemists have expended much
effort since the first revelations by the USDA Group[1] to
prepare materials which can be used cost-effectively in
absorbent products. We have seen the annual tonnage of
material used in such products increasing and we continue
to see new suppliers arriving in the marketplace with
another "super" product. Basically, it appears to me that
they differ only in degree of composition from the many
natural materials known as gums, exudates and extractives.

The mystery remains as to the reason why these particular organic polymer molecules are able to cause gross
association of water molecules. It is quite easy to calculate the number of the latter associated in various degrees
of immobility with each of the hydrogel molecules. As an
"explanation" for these actions, it is generally pointed
out that the useful compounds are:

a. hydrophilic

b. polyhydroxy

c. polycarboxy

but these "transparent" words do not explain the action; the terminology only serves to disguise our ignorance.

This lack of knowledge is very clearly seen when one evaluates "water-holding" capability in the presence of a small quantity of any electrolyte and compares it with the results obtained in the total absence of electrolytes. All of the synthetic materials and most of the natural materials exhibit this reproducible reaction to pure water versus electrolyte-containing liquids. Some of us have demonstrated that the reproducibility is truly remarkable, via alternate sequential exposure of crosslinked hydrogels to 1% saline and demineralized water. The change in swelling degree is not only remarkably reproducible but also remarkably rapid.

I believe that such a reproducible and simply demonstrated property should be examined for great potential expansion of applications of polysaccharides in industrial operations. Perhaps one starting point is through basic research on applications which have been very common for centuries and which actually produce the opposite effect; namely, the use of electrolytes (such as the calcium ion) to build rather than reduce viscosity in agar- and pectin-based hydrogels. We all enjoy our jellies and jam in the morning without thinking that there is a very valuable electrolyte-tolerant "superabsorbent."

REFERENCE

1. M.O. WEAVER, et al, Highly Absorbent Starch-Containing Polymeric Compositions, US 3,981,100, Sept. 21, 1976.

INVESTIGATIONS OF POLYSACCHARIDES IN THE SOLID STATE
BY ^{13}C CP-MAS NMR SPECTROSCOPY

Philip E. Pfeffer

U.S. Department of Agriculture

Eastern Regional Research Center

Philadelphia, Pennsylvania   19118

A brief introduction into the ^{13}C CP-MAS methodology
and a review of the recent application of this technique to the study of polysaccharides is presented.

1.   INTRODUCTION

For many years the NMR spectroscopist has been frustrated
with his inability to study the physical properties of
high molecular weight polymers such as polysaccharides
because of their insolubility. To circumvent these problems, often times the materials would have to be chemically modified, e.g., acetylated, hydrolyzed, or enzymatically cleaved to give smaller fragments. Inevitably
the substances derived from these various treatments bear
little or no resemblance in dynamic properties to the
original intact polymers. While X-ray crystallography has
been used extensively for characterizing such polymers as
cellulose, the interpretation of these results has been
less than unambiguous[1]. In addition, because of the time

constant associated with the crystallographic methodology
only marginal dynamic information can be obtained.

In contrast, high resolution ^{13}C solid state cross
polarization magic angle spinning NMR spectroscopy can be
used to examine the structures of bulk polysaccharide
polymers and derivatives in their native states. Measurement of relaxation processes can give information concerning
the intimacy (domain sizes) of coexisting polymers in such
a structurally diverse polymer as wood cellulose. In
addition this methodology can also pinpoint the origin of
crystalline and amorphous regions in the matrices as well
as establish the nature and asymmetry of the polymer
repeat units.

1.1   Solid State Methodology

Unlike the solution state, intermolecular interactions in
the solid state among the solute molecules not modulated
by the intervention of the solvent. In general such
interactions are dipolar in nature and manifest themselves
by giving rise to broad resonance lines in the ^{13}C NMR
spectra of these materials. Without the removal of these
interactions (primarily with protons), each ^{13}C resonance
line would be of the order of 2-3 KHz in width and each
spectrum would contain no discernible definition. To
alleviate this problem, high power decoupling has been
used (approximately 10 times the intensity required to
remove the scalar through-bond interactions in solution
^{13}C spectra) to achieve acceptable line narrowing[2].

Additional problems associated with obtaining high resolution ^{13}C spectra in the solid state result from the low sensitivity associated with the dilute ^{13}C nucleus. Also, there is a lack of intimacy of the ^{13}C nuclei in the solid, i.e., because of the restrictions imposed in the solid each ^{13}C nucleus cannot interact with one another as frequently as they do in solution and consequently cannot undergo mutual spin flips. The result is intolerably long spin lattice relaxation times, T_1, making data acquisition unfeasable under normal time constraints. To defeat this problem, a technique pioneered by Pines et al.[2], which exploits the large proton population, is used to facilitate relaxation on the rapid proton time scale. In essence this method uses a pulse sequence that induces the protons and carbon nuclei to precess at the same frequency in their respective rotating frames. During this period (milliseconds), which is the cross polarization time, the protons and carbons exchange energy through mutual spin flips and, thus, the rapidly relaxing proton population controls the carbon relaxation process. Typically, spin lattice relaxation times characteristic of cross polarization experiments are of the order of fractions of a second. Additionally the cross polarization process gives rise to ^{13}C signal enhancement due to polarization obtained from the proton population. This results in a fourfold increase in signal-to-noise over the noncross polarization experiment.

Finally, in order to remove a third critical factor responsible for the lack of sufficient resolution in the solid state, i.e., chemical shift anisotropy (CSA), the sample must be spun at frequencies as high as 3-4 KHz

(depending on the magnetic field used) at an angle of
54.7°C. Utilizing this technique with the two mentioned
above, ^{13}C cross polarization, magic angle spinning CP-MAS
NMR spectroscopy can produce relatively high resolution
spectra of crystalline solids and polymers in times comparable to those used to obtain solution spectra[3].

For a complete and detailed account of the CP-MAS
process and its application to the study of the structural
dynamics of polymers in general the reader is referred to
the extensive reviews by Yanoni[4], and Havens and Koenig[5],
respectively. More specific information on the applications
of ^{13}C CP-MAS NMR to carbohydrates can be found in a
recent review by Pfeffer[6].

1.2   Application to Polysaccharides

1.2.1   Studies of cellulose   The first CP-MAS ^{13}C studies
of cellulose I (native) and mercerized or regenerated
cellulose II clearly demonstrated that both crystalline as
well as amorphous regions could be defined[7,8]. One of
these studies[7] suggested that the doublet multiplicity of
the C-1 and C-4 resonances in cellulose I and II spectra
are indicative of unequivalent adjacent anhydroglucose
units ascribed to the basic repeat structural unit of
anhydrocellobiose. The other report suggested that this
splitting was due simply to two magnetically inequivalent
environments of glucose monomers[8]. Further refinement of
this work by Earl and VanderHart[9] showed that in samples
of cellulose I in the form of cotton, ramie, and hydro-
cellulose there was a clear splitting of the C-1 and C-4

resonances indicative of more than two anhydroglucose
residues per unit cell (possibly four) in the crystal
structure. In addition, these authors suggest that the
broad peaks observed in the spectrum do not originate from
paracrystalline regions of the sample, but are attributable
to anhydroglucoses on the surface of cellulose elementary
fibrils. This conclusion was based on the fact that these
broad resonances are missing in spectra of celluloses
obtained from Acetobacter xylinium and Valonia ventricosa
which have a very high degree of lateral organization and
low surface-to-volume ratio.

Maciel et al.[10] clearly addressed the question of
order and morphology of cellulose from a single source
that was modified mechanically. Their results clearly
supported those of Earl and VanderHart[9] that the broad
resonances associated with amorphous regions observed in
cellulose I samples from biological sources are primarily
due to cellulose monomers on the surface. An elegent
method for preferentially examining the crystalline or
amorphous regions of cellulose has been demonstrated by
Horii et al.[12]. This technique exploits the difference in
the spin-lattice relaxation times T_1^C in the laboratory
frame to suppress the spectrum of one domain while empha-
sizing the other. In addition, a quantitative assessment
of the ratio of amorphous to crystalline cellulose in
different cellulosic matrices was demonstrated following a
careful evaluation of the critical relaxation process
responsible for the spectral responses of each component.

In a detailed study Dudley et al.[11] examined the ^{13}C
CP-MAS spectra of different size cellulose oligomers to

determine at what stage the characteristic features of
cellulose II appears as the oligomer chain length was
increased. From cellotetraose through cellohexaose the
authors observed the characteristic C-1 resonance splitting reminiscent of cellulose II, as well as a convergence
of the C-1 and C-4 shift with those of cellulose II.
Careful integration of the C-1 doublet resonance in the
spectrum of cellotetraose demonstrated that these absorp-
tions were clearly in the ratio of 1:1. Since cellotetraose
contains only three interglycosidic linkages, alternation
of 0.4 dihedral angle pairs should lead to a 2:1 intensity
ratio for the C-1 doublet rather than the 1:1 ratio observed.
This observation was interpreted as evidence that supports
a structure of cellotetraose (and by extension cellulose II)
that necessitates two independent chains in the unit cell.
A most recent report by Atalla and VanderHart has refined
the CP-MAS spectra of native celluloses into two distinct
crystalline forms. The authors feel that the spectra of
native celluloses have multiplicities that cannot be
explained in terms of nonequivalent sites within a unique
unit cell[13] as described by Dudley et al.[11]. This work
proposed the existence of two different crystalline forms
Iα and Iβ which are in different proportions depending on
the source of cellulose, e.g., bacterial cellulose such as
Acetobacter cellulose containes 60-70% Iα, whereas plant
derived cotton contains 60-70% Iβ. These findings have
important implications in terms of understanding the
mechanism by which plants and microorganisms biosyntheti-
cally elaborate cellulose.

Horii et al.[14] have suggested that the chemical shift
position of the C-6 resonance in the CP-MAS ^{13}C spectra of
carbohydrates is diagnostic for the conformaton of the
CH_2OH groups about the exo-cyclic C-C bond. According to
X-ray analysis the conformations about the C-5-C-6 bond is
gauche-trans in α-D-glucose and gauche-gauche in α-D-
glucose·H_2O, where gauche-trans indicates that the C6-O6
bond is gauche to the C5-O5 bond and trans to the C4-C5
bond. It was therefore assumed for these samples that the
low field C-6 resonance at 64.5 ppm for α-D-glucose corre-
sponded to a gauche-trans conformation and the higher field
absorption for the C-6 resonance at 61.6 ppm for α-D-
glucose·H_2O to a gauche-gauche conformation. A good
linear correlation of C-6 shift position with torsion
angle χ around the exo-cyclic C-C bonds was obtained with
mono and disaccharides. An evaluation of native cellulose
in the crystalline state gave good agreement for its
assigned trans-gauche conformation. The position of
63.4-63.9 ppm for the C-6 resonance of noncrystalline
native cellulose suggest that the CH_2OH group is in the
gauche-trans conformation. Regenerated cellulose in the
crystalline forms gave a C-6 shift in the low field range
consistent with the gauche-trans conformation; however,
this interpretation is at odds with previous X-ray studies
that say the CH_2OH group exists in both gauche-trans and
trans-gauche conformations. The noncrystalline regenerated
cellulose has the same shift range and conformation as its
crystalline counterpart. Detailed studies[15] of ^{13}C chemical
shift assignments for α-D-glucose, α-D-glucose·H_2O, and
β-D-glucose, which exploit ^{13}C-^{13}C dipolar interactions in

specifically ^{13}C enriched glucose molecules have uncovered
differences in the corresponding ring carbon resonance
positions for α-$\underline{D}$-glucose and α-$\underline{D}$-glucose·H_2O that are
comparable to the C-6 shift differences mentioned above[14].
Such perturbations that presumably derive from intermolecular
interactions could also contribute significantly to the
C-6 shift differences for α-$\underline{D}$-glucose and α-$\underline{D}$-glucose·H_2O
attributed to purely conformational preferences. A compari-
son of the relative positions of the ring carbon resonances
in the solid state and those in solution for the glucose
series showed a poor correspondence, again indicating the
strong effect of crystal lattice interactions on chemical
shift[15]. For a complete review of the ^{13}C CPMAS work
being done in the area of cellulose chemistry see the
article by Fyfe et al.[16].

1.2.2   Studies of Other Polysaccharides

^{13}C solid state methodology is becoming useful for determining the primary structure of various polysaccharides
and their derivatives[17]. This is a particularily important
application for derivatives such as xanthan gum and substituted chitosans. Conformational information in (1-3)-β-$\underline{D}$-
glucans in both noncrystalline and crystalline regions has
also been gleaned from its ^{13}C spectra[18]. Following a
study of the ^{13}C chemical shifts of various polysaccharides
(curdlan, lentinan, and laminaran) in the solid, gel, and
solution it was concluded that the high molecular weight
glucans adopt mainly a helix form, while the low molecular

weight assume a random coil form in addition to the helix in the solid.

A preliminary study comparing the ^{13}C solid state spectra of various amyloses of differing degrees of poly-merization (DP) values and treatments showed that increased crystallinity significantly narrowed the inherent line-widths[19]. In addition, these authors observed a 3 ppm up-field shift of the C-1 ^{13}C resonance in the lower molecular weight amylose (DP 25) when compared with the higher DP 100 sample. This difference in C-1 shift was attributed to differences in conformational populations in the two samples.

1.3 Complex Polysaccharides Matrices

Lodgepole Pine wood was subjected to various mechanical (grinding, ball melting) and chemical processing and the fractions then examined by ^{13}C CP-MAS NMR[20]. Both lignin and carbohydrate fractions gave spectra reflecting their characteristic compositions. The spectra gave indirect evidence for the existence of lignin-carbohydrate complexes since all lignin fractions exhibited carbohydrate signals and vice versa. Spectra of explosion-treated spruce wood suggest that the explosion treatment gives increased crystallinity to the cellulose since their signals are sharper than those obtained from untreated samples[21]. Exploded wood pulp has also been examined by a number of pulse sequences in order to more clearly define the nature of internal lignin-carbohydrate complexes[22]. In addition to quantifying the lignin[23] in this material the criteria of homogeneous proton spin diffusion was used to characterize a single-phase lignin-carbohydrate complex residue derived from cellulase treatment of the pulp[22]. In addition doping of the intact wood pulp with paramagnetic Fe^{+3} gave a clear indication that rapid spin diffusion could be efficiently transferred from the Fe^{+3} bound to the carbohydrate to the lignin component[22]. ^{13}C CP-MAS has also been effective for monitoring polysaccharide breakdown or disordering in ripening apple cell wall tissue[24]. In this study critical point dried cell walls were examined via various spin lattice relaxation times to estimate changes in polymer mobilities as a function of ripening. Of particular interest were the carbonyl resonances of the polyuronide which showed a 42% drop in proton T_1 over a period of 21 days. These data correlate well with a decrease in fruit tissue firmness over this period of time, indicating that the pectic substances ("glue") within the matrix is breaking down and becoming increasingly disordered[21]. Sodium T_1 values for this tissue show the same trend. Parallel model experiments using cell wall tissue treated with endo-cleaving polygalacturonase enzyme PG II also demonstrated a 57% drop in proton T_1 values for the carbonyl resonance, indicating the effect of polyuronide chain shortening on overall relaxation rates.

REFERENCES

1. a) T. PETITPAS, M. OBERLIN, and J. MERING, J. Polymer Sci., S2, 423 (1963); b) M. NORMAN, Text. Res. J., 33, 711 (1963); c) A. SARKO, R. MUGGLI, Macromol., 7, 486 (1974)

2. A. PINES, M. G. GIBBY, and J. S. WAUGH, J. Chem. Phys.,
    56, 569 (1973)
3. J. SCHAEFER, E. O. STEJSKAL, and R. BUCHDAHL, Macromol.,
    10, 384 (1977)
4. C. S. YANONI, Acct. Chem. Res., 15, 201 (1982)
5. J. R. HAVENS and J. L. KOENNIG, Appl. Spectrosc., 37,
    226 (1983)
6. P. E. PFEFFER, J. Carbohyd. Chem., in press (1984)
7. R. A. AHALA, J. C. GAST, D. W. SINDORF, V. J. BARTUSKA,
    and G. E. MACIEL, J. Am. Chem. Soc., 102, 3249 (1980)
8. W. L. EARL and D. L. VANDERHART, J. Am. Chem. Soc.,
    102, 3251 (1980)
9. W. L. EARL and D. L. VANDERHART, Macromol., 14, 570
    (1981)
10. G. E. MACIEL, W. L. KOLODZIEPKI, M. S. BERTON, and
    B. E. DALE, Macromol., 15, 686 (1982)
11. R. L. DUDLEY, C. A. FYFE, P. J. STEPHENSON, Y.
    DESLANDES, G. K. HANER, and R. H. MARCHESSAULT,
    J. Am. Chem. Soc., 105, 2469 (1983)
12. F. HORII, A. HIRAI, and R. KITAMARU, J. Carbohyd. Chem.,
    in press (1984)
13. R. H. ATALLA and D. L. VANDERHART, Science, 223, 284
    (1984)
14. F. HORII, A. HIRAI, and R. KITAMARU, Polymer Bull.,
    10, 357 (1983)
15. a) P. E. PFEFFER, K. B. HICKS, M. H. FREY, S. J.
    OPELLA, and W. L. EARL, J. Mag. Res., 55, 344 (1983);
    b) P. E. PFEFFER, K. B. HICKS, M. H. FREY, S. J.
    OPELLA, and W. L. EARL, J. Carbohyd. Chem., 3, 197
    (1984)
16. C. A. FYFE, R. L. DUDLEY, P. J. STEPHENSON, Y.
    DESLANDES, G. K. HAMER, and R. H. MARCHESSAULT,
    Rev. Macromol. Chem. Phys., 23, 187 (1983)
17. L. D. HALL and M. YALPANI, Carbohydr. Res., 91 C1
    (1981)
18. H. SAITO, R. TABETA, and T. HARADA, Chem. Lett., 571,
    (1981)
19. H. SAITO and R. TABETA, Chem. Lett., 713, (1981)
20. W. KOLODZIEJSKI, J. S. FRYE, and G. E. MACIEL,
    Anal. Chem., 54, 1419 (1982)
21. M. G. TAYLOR, Y. DESLANDES, T. BLUHM, R. H.
    MARCHESSAULT, M. VINCENDON, and J. SAINT-GERMAIN,
    Tappi J., 66, 92 (1983)
22. W. V. GERASIMOWICZ, K. B. HICKS, and P. E. PFEFFER,
    Macromol., in press (1984)

23. J. F. HAW, G. E. MACIEL, and H. A. SCHROEDER,
     Anal. Chem., in press (1984)
24. P. L. IRWIN, P. E. PFEFFER, W. V. GERASIMOWICZ, R.
     PRESSEY, and C. E. SAMS, Phytochem., 23, (1984)

AN ALTERNATIVE CHIROPTICAL TECHNIQUE FOR THE STUDY OF IONIC-POLYSACCHARIDE SOLUTIONS

V. CRESCENZI
Department of Chemistry & Chemical Engineering
Stevens Institute of Technology
Castle Point, Hoboken, NJ 07030

H. G. Brittain
Department of Chemistry
Seton Hall University
South Orange, NJ 07079

Y. Okamoto
Department of Chemistry
Polytechnic Institute of New York
Brooklyn, NY 11201

We wish to report some results which have been recently obtained in our laboratories, and which deal with the application of circularly polarized luminescence (CPL) spectroscopy to the study of ionic polysaccharides in aqueous solution. To introduce the matter, it is first necessary to briefly describe the following considerations.

While circular dichroism (CD) spectroscopy in the ultraviolet and visible regions of the spectrum has received the largest degree of attention, an entire series of alternative chiroptical techniques is in the process of development at the present time. The optical activity associated with molecular vibrations can be measured by the vibrational-CD technique, and the use of FTIR-VCD methods[1] should enable the method to be used with carbohydrates.

Similarly, molecules having an appropriate fluorophore may
be excited with circularly polarized light, and the resulting
fluorescence studied as a function of excitation wavelength.
This technique has been termed fluorescence detected circular
dichroism, or FDCD.[2] Finally, another chiroptical technique
has been developed which involves the measurement of the
differential emission of left- and right-circularly polarized
light emitted by a chiral luminescent compound. This latter
effect has been termed circularly polarized luminescence.[3]

The CPL technique yields information on the molecular
chirality of the luminescent excited state in exactly the
same manner as CD spectroscopy reflects the molecular
chirality of the ground state. If the geometry of the
molecule remains invariant during the excitation and emission
processes, then CD and CPL spectroscopies will yield the
same results regarding the optical activity of the species
under study. Should a geometrical change accompany the
excitation process, then the CD and CPL would be quite
different.

The sole restriction placed on CPL spectroscopy is
that it is confined to luminescent species only. This
situation actually insures that the CPL technique is highly
selective in nature, since one obtains the chirality of only
the emitting portion of the molecule. In addition, should
the absorptivity of a molecule be too weak as to permit
the use of reasonable concentrations for CD measurements
(such as in the f-f transitions of lanthanide complexes,
or for singlet-triplet transitions of organic molecules),
the CPL technique may be the only method by which chiroptical
information can be obtained.

Probably the largest number of CPL studies which have

been carried out to date have dealt with chiral metal complexes, and essentially all of these studies have been concerned with the f-f optical activity of lanthanide complexes.[3b] Of all the lanthanide ions, Tb(III) and Eu(III) are found to emit with the highest intensity, and these are observed to luminesce in a series of well-defined bands even in fluid solution at room temperature. CPL spectra have been obtained for many different systems, including a number of biopolymers. For instance, it has been established that Tb(III) and Eu(III) function as excellent luminescence probes for the study of proteins[4] and nucleic acids.[5] In addition, it has recently been reported that the intensity of luminescence emitted by these lanthanide ions can be strongly enhanced when these are bound by synthetic poly-electrolytes (such as sodium polyacrylate) in dilute aqueous solutions.[6]

It was therefore considered of great interest to study the luminescence of Tb(III) ions in dilute solutions of ionic polysaccharides, inasmuch as the latter are truely chiral species. The binding sites on the biopolymers should chelate lanthanide ions with great efficiency, giving rise to CPL spectra through a variety of established mechanisms.[3]

Using the CPL apparatus built at the Department of Chemistry at Seton Hall University,[7] we have found that a number of different ionic polysaccharides strongly bind Tb(III) in dilute aqueous solutions. In the case of carboxylated polysaccharides, one also obtains characteristic CPL spectra[8] in addition to reasonably strong luminescence.

The results collected thus far have enabled the following general trends to be deduced:

1) The total luminescence intensity emitted by the lanthanide

ions can be greatly enhanced upon binding to polysaccharide polyelectrolytes. The emission enhancement
depends (for a given common polymer/counterion equivalent
concentration and total ionic strength) on the nature
and density of fixed charges along the polysaccharide
chains.

2) While formation of Tb(III)/polysaccharide complexes
invariably leads to enhanced Tb(III) luminescence, CPL
is not invariably observed.   In the cases studied to
date (heparin sulfate, various carboxymethyl celluloses,
alginate, polygalacturonic acid, sclerox S-1.0, and
xanthan), the general trend is that the presence of
carboxylate groups along the polysaccharide chains is
required to produce the well defined, chiral binding
site which leads to observable CPL. The only exception
found thus far consists of xanthan in the "native" form:
in this case, no Tb(III) CPL spectrum could be obtained.

3) The CPL spectra of Tb(III) complexes with carboxylated
polysaccharides were found not only to depend on the
identity of the biopolymer, but for each given species
also depend on the chain conformation and the state of
aggregation associated with the polymer chains.

The CPL technique can therefore be considered as an
alternative chiroptical technique, useful for probing the
geometry and configuration of binding sites along carbohydrate polyelectrolytes and for studying the conformational
state of the chains.

The immediate goals of our research are to to obtain
a comprehensive picture of the dependence of lanthanide ion
CPL upon polysaccharide chemical constitution, structure,
and solution conformation. We seek to determine the possible

mechanisms whereby the Tb(III) ions experience the chirality of their environment, and seek to determine the degree of contribution made by vicinal, conformational, and configurational effects.

Quite naturally, once a sufficient number of cases have been examined in detail, we will derive a series of quantitative CPL spectra-structure correlations for the classes of biopolymers and derivatives of our concern. To this end, the full potentiality of the CPL technique will be exploited by studying the photophysical properties of polysaccharide chains (including the non-ionic polymers) labeled with different fluorescent probes.

REFERENCES

1. (a) L.A. Nafie, M. Diem, D.W. Vidrine, J. Am. Chem. Soc., 101, 496 (1979).
   (b) E.D. Lipp, C.G. Zimba, L.A. Nafie, Chem. Phys. Lett., 90, 1 (1982).

2. (a) D.H. Turner, I, Tinoco, Jr., M. Maestre, J. Am. Chem. Soc., 96, 4340 (1974).
   (b) I. Tinoco, Jr., D.H. Turner, J. Am. Chem. Soc., 98, 6453 (1976).

3. (a) F.S. Richardson, J.P. Riehl, Chem. Rev., 77, 773 (1977).
   (b) H.G. Brittain, Coord. Chem. Rev., 48, 243 (1983).

4. (a) R.B. Martin, F.S. Richardson, Quart. Rev. Biophys., 12, 181 (1979).
   (b) W.D. Horrocks, Jr., D.R. Sudnick, Acc. Chem. Res., 14, 384 (1981).

5. J. Eisinger, A.A. Lamola, Biochim. Biophys. Acta, 240, 299 (1971).

6. I. Nagata, Y. Okamoto, Macromolecules, 16, 749 (1983).

7. H.G. Brittain, J. Am. Chem. Soc., 102, 3693 (1980).

8. V. Crescenzi, H.G. Brittain, Y. Okamoto, N. Yoshino, J. Polymer Sci., Polymer Phys. Ed., in press.

RECONCILIATION OF THE DIFFERENCES IN NUMBER AVERAGE
MOLECULAR WEIGHTS ($\bar{M}_n$) BY END GROUP TITRATION AND
OSMOMETRY FOR LOW METHOXY PECTINS

M. L. FISHMAN AND L. PEPPER

Eastern Regional Research Center
Agricultural Research Service
U.S. Department of Agriculture
Philadelphia, Pennsylvania 19118

Membrane osmometry on protonated low methoxy pectins
(degree of methylation, 35 and 37) revealed that
these macromolecules formed aggregates which dissociate
slowly over a period of days when activated by heat.
Plots of π/c against c for these pectins show broad
maxima when the system is sufficiently removed from
equilibrium. At or near equilibrium, the van't Hoff
plot resembles conventional dissociating macromolecular systems and in the lim c$\rightarrow$o (π/c), measured
π/c approach the values calculated from $\bar{M}_n$ obtained
from end group titration.

1. INTRODUCTION

Recently[1] for various pectins, we have found that $\bar{M}_n$ from
membrane osmometry (o) was about two to four times greater
than the value from end group titration (EGT). We suggested
that $\bar{M}_{n(o)}$ was the aggregated molecular weight in solution
for pectin whereas $\bar{M}_{n(EGT)}$ was the value for disaggregated
or pectin "monomer." Furthermore,[2] plots of reduced
osmotic pressure (π/c) against concentration (c) for
protonated and sodium salts alike pass through a minimum.
From these results, it was concluded that pectins are

nonideal aggregating systems. Moreover, kinetic studies
indicated that these systems, when first dissolved, formed
aggregates which dissociated slowly over days after briefly
heating. In the case of the protonated pectins, the
medium methoxy pectins (degrees of methylation, DM 57 and
58-60) exhibited approximately 3 days lag time between
heating and onset of dissociation. The protonated high
methoxy pectins DM 70 and 72-73, exhibited a lag time of
about 5 days, whereas the protonated low methoxy pectins
(DM 35 and 37) had not dissociated after 12 days.[3]

In this note, we extend our studies on the dissociation of the protonated DM 35 and 37 pectins in 0.047 M
NaCl and 0.003 M NaN_3.

2. EXPERIMENTAL

Commercial citrus pectin 35% degree of methyl esterification (DM) was a gift from Bulmers Limited,* Hereford,
England. Another pectin, also manufactured by Bulmers,
DM 37 was a gift from Drs. E. R. Morris and M. J. Gidley
at Unilever. Characterization of the pectins have been
described previously.[4] Pectin (2 g) was dissolved in
60 mL of distilled water with gentle stirring overnight at
5°C, dialyzed against four changes of water over a 48-h
period in a Spectrapor bag with a 12,000 molecular weight
cut-off, centrifuged 1 h at 30,000 g and lyophillized. At
the beginning of dialysis, the outside-inside volume ratio

---

* Reference to brand or firm name does not constitute
endorsement by the U.S. Department of Agriculture over
others of a similar nature not mentioned.

was 30:1. Osmotic pressure at 35.0 ± 0.1°C was measured
as before.[5]

2.1. Approach to Equilibrium

Samples were allowed to approach equilibrium in two ways.
Typically, 200 mg samples were dissolved in 20 mL of salt
solution and placed in capped glass bottles. Then samples
were emersed in boiling water for 10 min, followed by
equilibration at 35 ± 1°C.

2.1.1 Procedure 1

In this procedure, samples were equilibrated for 3 to
29 days at a concentration of 1 g/dl and diluted serially
just prior to measurement of osmotic pressure.

2.1.2 Procedure 2

After cooling to 35°C, the samples were diluted immediately
and allowed to stand for 22 or 29 days prior to measuring
osmotic pressure.

3. van't Hoff Plots

The results of these experiments are displayed in Figures
1 and 2. The reduced osmotic pressure, π/c, is plotted
against concentration (c), according to the van't Hoff
limiting law.[6]

$$\pi/c = RT[1/\overline{M}_n + Bc] \tag{1}$$

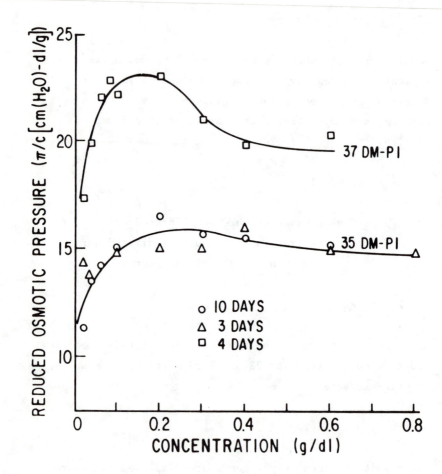

FIGURE 1   van't Hoff plots after 3 and 10 days for 35%
            degree of methylation pectin (DM) and 4 days
            for DM 37%.   Approach to equilibrium by
            procedure 1.

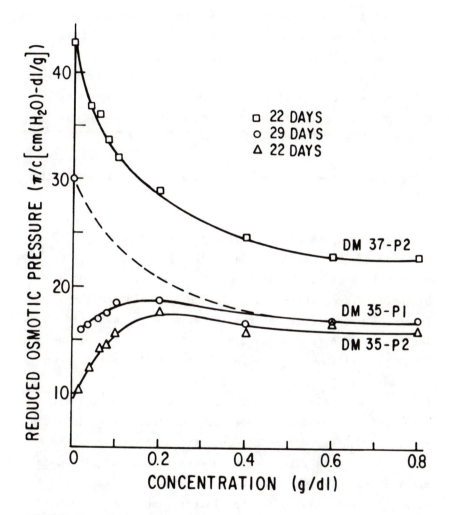

FIGURE 2   van't Hoff plots after 22 and 29 days for 35% degree of methylation pectin and 22 days for DM 37%. Approach to equilibrium by procedure 1 for DM 35, 29 days, and by procedure 2 for others.

For a polydisperse system, the intercept of such a plot
yields the number average molecular weight, $\bar{M}_n$, whereas
the slope yields the second virial coefficient, B.

The data in Figure 1 were results from a procedure 1
(P1) approach to equilibrium. After 10 days, there is
little if any change in osmotic pressure at any concentration for DM 35. These results confirm our previous conclusion that low methoxy pectin aggregates are relatively
stable over a 10-12 day period.[2] A similar curve but with
a higher maximum was obtained for DM 37 after 4 days.

In Figure 2 are the results of a 22 day procedure 2
(P2) experiment on DM 37, a 29 day P1 experiment on DM 35
and a 22 day P2 experiment on DM 37. Interestingly, the
van't Hoff plot of the DM 35 P2 experiment is that of a
classical dissociating system.[7] The π/c value at the
intercept was calculated from the value of $\bar{M}_n$ obtained
independently by end group analysis.[2] Comparison of the
DM 37 P2 experiment in Figure 2 with the P1 experiment on
the same polymer in Figure 1, reveals that when c >
0.4 g/dl, π/c values are much closer to equilibrium values
than when c < 0.4 g/dl. Moreover, the differences in π/c
increase with decreasing concentration. A comparison of
the DM 35 P1 data after 3 and 10 days (Figure 1) with data
after 29 days (Figure 2), reveals that π/c has increased
at all concentrations, for the latter case, but the maximum persists. In the case of the DM 35 P1, had π/c values
at low concentration followed the dashed line, the van't
Hoff plot would not have been unusual. The only known
significant difference between these two is that DM 37 has
roughly 65% the $\bar{M}_n$ of DM 35 by end group titration.

Furthermore, osmometry on the sodium salts reveals that
DM 37 aggregates have three-quarters the $\bar{M}_n$ found for the
DM 35 aggregates.[2]

4. CONCLUSIONS

The foregoing lead to several inferences and conclusions
regarding low methoxy protonated pectin aggregates.

1. They form metastable aggregates which are more
stable than all other pectins with higher DM or than
neutralized pectins of similar DM.

2. Their state in solution and rate of approach to
equilibrium are concentration dependent, which could
account for maxima in van't Hoff plots.

3. The concentration dependence of aggregation is
more marked at higher concentrations for DM 35 or 37
protonated pectins than for all other pectins with higher
DM or than neutralized pectins of similar DM. Thus,
curves may be pseudo ideal at higher concentrations due to
a balancing of dissociation effects and positive second
virial effects.

4. At or near equilibrium, lim c→o (π/c) osmometry
approaches π/c calculated from end group titration.

REFERENCES

1. M. L. FISHMAN, L. PEPPER, P. E. PFEFFER, R. A.
   BARFORD, Pectin Aggregation as Measured by Number
   Average Molecular Weights, Abstracts of Papers, 186th
   ACS Natl. Meeting, Washington, DC, AGFD (1983).
2. M. L. FISHMAN, L. PEPPER, P. E. PFEFFER, Dilute
   Solution Properties of Pectin, Preprints of Papers

DMSE Division, 188th ACS Natl. Meeting, Philadelphia, PA (1984).

3. M. L. FISHMAN, L. PEPPER, P. E. PFEFFER, in Characterization of Water Soluble Polymers, edited by J. E. Glass (American Chemical Society), In Press.

4. M. L. FISHMAN, P. E. PFEFFER, R. A. BARFORD, L. W. DONER, J. Agr. Food Chem., 32, 372 (1984).

5. M. L. FISHMAN, L. PEPPER, R. A. BARFORD, J. Poly. Sci. Physic. Ed., 22, 899 (1984).

6. C. TANFORD, Physical Chemistry of Macromolecules, Chapter 4 (John Wiley & Sons, New York, 1961).

7. M. P. TOMBS, A. R. PEACOCKE, The Osmotic Pressure of Biological Macromolecules, Chapter 4 (Clarendon Press, Oxford, England, 1974).

SMALL ANGLE X-RAY SCATTERING FROM POLYSACCHARIDE SOLUTIONS

S.S. STIVALA
Department of Chemistry and Chemical Engineering
Stevens Institute of Technology
Hoboken, New Jersey 07030

Certain structural features, e.g., mass, size, shape, of macromolecules in solution, not readily elucidated from conventional methods such as light scattering or ultracentrifugation, may be obtained from small angle x-ray scattering, SAXS. A point in question is our earlier work on heparin[1]. Though the molecular, weight, M, and the sedimentation coefficient, $S°$, in water of this important glycosaminoglycan, were reported by us and other investigators (see references 2 and 3 for key references) no information was available on the radius of gyration, R_g. Though, in principle, the mass of heparin could be assessed from light scattering its R_g cannot be since the molecule is much too small relative to the wavelength, λ, of visible light (about 3500 to 5400 Å) to exhibit disymmetry of light scattering. On the other hand, macromolecules are large compared to the wavelength of x-rays ($\lambda = 1.54$ Å from CuK_α). Our investigation of a heparin fraction from SAXS,[1] under conditions where polyelectrolyte behavior is minimized, yielded the following information. The M = 12,900 which was in excellent agreement with the values 12,600 observed

from sedimentation and 12,500 measured from intrinsic viscosity. The mass per unit length, M_u, derived from the
absolute intensity, was 54.5 Daltons per $\overset{o}{A}$ compared to the
calculated value of 52.7 on the basis of its structure
where the length of the monomer is identical to half the
cell length (5.15 $\overset{o}{A}$). It may be concluded from these observations that heparin is unbranched. The total length
of the heparin molecule, L, was obtained as 237 $\overset{o}{A}$ compared
to the theoretical calculated value of 245 $\overset{o}{A}$. The radius
of gyration of the cross section R_q, was obtained from
SAXS as 4.6 $\overset{o}{A}$ and the persistence length, a^*, was observed
to have the value of 21.1 $\overset{o}{A}$. Under the conditions in
which heparin was examined its behavior is described as a
Gaussian coil. The R_g was found to be 35.0 $\overset{o}{A}$.

In scattering phenomena the scattering angles are
inversely related to the dimensions of the scattering
particles. In the case of dilute monodisperse solutions
of macromolecules the scattered intensities from individual
molecules are additive, hence the scattering curve
(scattered intensity, I, *versus* scattering angle, 2θ)
provides information on the individual particle. X-rays
are primarily scattered by electrons and, therefore, SAXS
is observed when electron density inhomogenieties of
colloidal size exist in the sample. The experimental
method and theory are well established and may be found
in the recent book by Glatter and Kratky.[4] SAXS can provide, under certain conditions and in some instances,
structural parameters in addition to those mentioned in
our case for heparin, e.g., area, volume, thickness and
solvation.

The structural parameters derived for heparin are

based on the theory of SAXS applicable to thin long linear
molecules. Several years ago we were interested in assess-
ing structural features of branched polysaccharides in
solution, such as dextran and levan, from SAXS. We have
found that the same relationships of SAXS used for unbranched polymers were applicable to comb branch structures. This was based on studies of carefully characterized synthesized model polystyrenes.[5] Several samples of
hydrolyzed fractionated dextran, produced from
L. *mesenteroides*, were examined in aqueous solutions.[6,7]
The molecular weights from SAXS were found to be in very
good agreement to those obtained from light scattering, LS.
Table 1 compares values of M and R_g for the dextran
hydrolyzates obtained from SAXS and LS.

TABLE I   Solution Parameters of Dextran Hydrolyzates

M(SAXS)	246,000	105,000	74,500	42,000	12,000
M(LS)	253,000	106,900	72,500	44,900	11,200
R_g(SAXS),$\overset{o}{A}$	150	109	82	68	38
R_g(LS),$\overset{o}{A}$	139	98	82	65	35
R_q, $\overset{o}{A}$	6.05	6.10	4.32	2.65	2.00
L, $\overset{o}{A}$	4200	2490	1700	1100	355
M_u,$\overset{o}{A}{}^{-1}$	58.57	42.17	43.82	37.84	33.84

The mass per unit length for the unbranched dextran
may be computed from the monomer molecular weight (162
Daltons)and from its length (5.15 $\overset{o}{A}$). If branching is
to exist then the experimental M_u should exceed this
value. That this is, indeed, the case may be noted from
the data in Table I, where M_u for the dextran hydroly-

zates increases from 33.80 to 58.57 Daltons per $\overset{o}{A}$, with increasing molecular weight. The parameter L is the length of the backbone of the branched dextran.

In a similar manner the SAXS data of two previously characterized *S. salivarius* levan fractions were analyzed.[8,9] The results are summarized in Table II.

TABLE II   SAXS Data of *S. salivarius* Levan in H_2O at 25°C

Parameter	E-7	F-14
M	26,750	16.9×10^6
$R_{g,\overset{o}{A}}$	47.6	395
$R_{q,\overset{o}{A}}$	8.2	22.0
M_u, per $\overset{o}{A}$	45.8	72.9
L, $\overset{o}{A}$	534	2.3×10^5

The values of M and R_g from SAXS are in good agreement to those obtained from light scattering. The $(M_u)_1$ for a linear levan is computed from M_o/h where the M of the monomer, M_o, and monomer length, h, of 162 Daltons and 5.87 $\overset{o}{A}$, respectively, yielded the value of $(M_u)_1$ of 28.94 Daltons/$\overset{o}{A}$. That the two levan fractions are branched are confirmed from their values of M_u which exceed that of $(M_u)_1$.

Recently, we have turned our attention in examining solution parameters of proteoglycans, specifically that derived from bovine nasal cartilage. This effort is in collaboration with Professor John Gregory and Dr. S. Damle of the Rockefeller University in New York City. Proteoglycans are major constituents of connective tissues and

are found in high concentrations in blood vessels, skin,
joint fluid, cartilage, the vitreous body of the eye, and
in other locations. It consists of a protein or poly
(amino acid) chain essentially composed of one hundred or
more amino acid units and which can be a random arrangement of any of the twenty or so naturally occurring L-α
aminoacids. The protein chain forms the backbone of the
molecule with the carbohydrate moiety of the molecule
taking the form of polysaccharide chains as pendant groups
covalently bound to the protein chain. These polysaccharide chains are linear and fairly regular, possessing
alternating monosaccharide sequences. The polysaccharide
chains are strongly acidic with the major carbohydrate
constituent consisting of chondroitin 4-sulfate. The
proteoglycans of bovine nasal septum exist *in vivo* as very
large complex aggregates which can be dissociated at high
concentrations of guanidine hydrochloride into a mixture
of proteins and proteoglycan subunit molecules.[10] The
proteoglycan subunit is, in effect, a branched structure.

Samples of bovine nasal proteoglycan subunits were
prepared by Professor Gregory and Dr. Damle from which we
have obtained SAXS data in 0.15 M LiCl and in water.[11]
The values of M measured in 0.15 M LiCl and in H_2O were
2.25×10^6 and 2.55×10^6, respectively, with a value of
R_g = 494 Å in 0.15 LiCl. The intrinsic viscosities, [η],
in 0.15 LiCl, 0.15 M guanidium hydrochloride, GHCl, and 4.0
M GHCl were 0.098 dl g^{-1}, 0.603 dl^{-1} and 0.086 dl g^{-1},
respectively. The values of M and R_g obtained from
SAXS are in good agreement with 2.5×10^6 from sedimentation equilibrium[10] and 570 A from light scattering[12] in 4.0
M GHCL, respectively. The [η] that we measured in 0.15 M

LiCl and in 4.0 M GHCl are in the same order of magnitude, hence comparison of R_g values in 0.15 M LiCl from SAXS to that in 4.0 GHCl from LS is justified. The processing of our SAXS data of the proteoglycan in 0.15 M LiCl and in H_2O to include values of R_q, M_u, a^*, is currently near completion and will be subsequently published. Contrast variation studies should enable the elucidation of structural features of the protein as differentiated from that of its polysaccharide branches. This may be effected by matching the electron density of the solvent (H_2O and sucrose) to that of either the protein, or polysaccharide, thus obtaining scattered intensities from the unmasked moiety.

REFERENCES

1. S.S. STIVALA, M. HERBST, O. KRATKY, and I. PILZ, Arch. Biochem. Biophysics, 127, 795-802 (1968).
2. J. EHRLICH and S.S. STIVALA, J. Pharm. Sci., 62, 517-544 (1973).
3. S.S. STIVALA, Federation Proc., 1, 83-88 (1977).
4. O. GLATTER and O. KRATKY, Small Angle X-Ray Scattering (Academic Press, New York, 1982).
5. S.K. GARG and S.S. STIVALA, Polymer, 23, 514-520 (1982).
6. S.K. GARG and S.S. STIVALA, J. Polym. Sci. Phys. Ed., 16, 1419-1434 (1978).
7. S.K. GARG and S.S. STIVALA, J. Polym. Sci.: Polym. Phys. Ed., 18, 405-407 (1980).
8. S.S. STIVALA and B.H. KHORRAMIAN, Carbohydrate Res., 101, 1-11 (1982).
9. B.H. KHORRAMIAN and S.S. STIVALA, Carbohydrate Res., 108, 1-11 (1982).
10. V.C. HASCALL, and S.W. SAJDERA, J. Biol. Chem. 244, 2384 (1969); 245, 4920-4930 (1970).
11. To be published by S.S. STIVALA, A. PATEL, J. GREGORY and S. DAMLE.
12. S.G. PASTERNACK, and A. VEIS, J. Biol. Chem. 249, 2206-2211 (1974).

SELECTIVE FUNCTIONALIZATION OF POLYSACCHARIDES

DEREK HORTON

Department of Chemistry

The Ohio State University

Columbus, Ohio 43210, USA

With major emphasis on the (1→4)-D-glucans amylose and cellulose, this report focuses on substitutive, oxidative, and deoxygenative modifications of natural glycans. Selective substitution provides entry into modifications selectively oxidized at C-2 or C-3 and thence, by way of oxime derivatives, into specifically aminated analogs (cationic polysaccharides). Indirect oxidation at C-6 by the sequence 6-chlorination ⟶ azide displacement ⟶ photolytic conversion (through an imine) into the corresponding 6-aldehyde is highly specific and can be conducted at various d.s. levels. The position of oxidation in variously oxidized polysaccharide derivatives may be monitored conveniently by carbon-13 n.m.r. spectroscopy; complementary analyses may be made by specific deuteration through reduction with borodeuteride, with subsequent mass-spectrometric determination of label position in the monomer unit. Eliminative 2,3-deoxygenation provides highly reactive, 2,3-unsaturated polymers. Deoxygenation at C-6 may be achieved by photolysis of 6-iodo derivatives of polysaccharides.

1. INTRODUCTION

Work in our laboratory concerned with chemical
structural interconversions of sugars and their
derivatives has had a sustained component addressing
the question of such structural modifications at the
polysaccharide level. Such transformations may be
broadly classified into O-substitution reactions,
deoxygenation reactions, and oxidative
transformations. All three of these aspects are
addressed in this report, with major emphasis on the
$(1 \rightarrow 4)$-D-glucans, namely amylose and cellulose.

2. GENERAL

Although the chemical transformations of
monosaccharide sugars provide a working basis for
attempting similar transformations at the
polysaccharide level, significant differences exist,
and reactions feasible with monosaccharides and their
derivatives are not necessarily directly realizable
with polysaccharides. The effect of structural
organization beyond the primary level, accessibility
by reagents, and reactions involving more than one
sugar residue, are some of the factors that
complicate the picture in chemical modification of
polysaccharides. In any chemical transformation at
the polysaccharide level, an important question
concerns the extent to which such a reaction is
regular and uniform throughout the polysaccharide
chain, in contrast to block reactions in which some
regions of the chain are highly substituted while

other regions are modified to a lower extent than the statistical average for the entire structure.

Average structures may be determined by calculation[1] from the results of elemental analysis, but additional insight into the detailed structure requires such techniques as depolymerization and characterization of individual component residues, and importantly the use of such physical methods as proton and carbon-13 n.m.r. spectroscopy for probing into structural modifications on the intact polymer to assess the type and extent of substitution and its degree of regularity or randomness. Specific enzymes may also be of considerable utility in this regard.

In addition to the characterization of modified structural units, other important criteria involve the extent of depolymerization and/or crosslinking. The controlled modification of the properties of polysaccharides as a function of chemical derivatization, frequently at very low levels of transformation (degree of substitution, d.s.), is technologically significant.

A large body of literature has accumulated over many years on chemical transformations of polysaccharides. Much older work was conducted at a time when methods for characterization were inadequate to determine detailed structural consequences of reactions performed under empirical conditions. With the advent of powerful new techniques, many products described in earlier literature could be fruitfully reinvestigated.

3.   SELECTIVE O-SUBSTITUTION

Selective reaction of the primary position in amylose
or cellulose offers access not only to selectively O-
substituted derivatives but also to a range of other
derivatives, including oxidized functionalities
introduced by an indirect procedure that does not
require direct oxidation. It has long been known[2]
that some measure of selectivity for the primary
position may be achieved in the tosylation of amylose
if the degree of substitution by tosyl groups is kept
relatively low, but the selectivity for the primary
position decreases as the extent of tosyl
substitution increases. Thus treatment of pyridine-
swollen amylose with p-toluenesulfonyl chloride in
pyridine, employing slightly less than one mole of
reagent per glucose residue, followed by acetylation,
gives a product[3] whose total degree of substitution
(d.s.) by glucosyl groups is indicated to be 0.85 by
elemental analysis (Fig. 1).

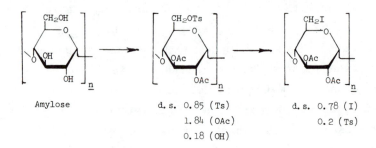

FIGURE 1   Tosylation of Amylose

The proportion of tosyl substitution at the primary position may be determined by reaction with sodium iodide, and the extent of replacement by iodide (which is specific[4] for the primary position under the conditions utilized) assures that while most of the substitution by tosyl groups is at the primary position, a small proportion of secondary sulfonate also is present[3]. Attempts to increase the substitution at the primary position by use of additional reagents leads to further reaction at the secondary positions.

3.1. Indirect Oxidation at C-6

Controlled tosylation of amylose to d.s. ~0.6 (Fig. 2) gives material substituted exclusively at primary positions.

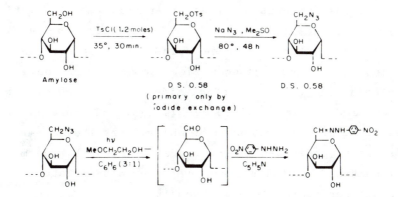

FIGURE 2   Reactions of 6-Azido-6-deoxyamylose

These primary tosyloxy groups may be completely
replaced by azide in a displacement reaction
conducted in dimethyl sulfoxide, thus providing
access via reduction to a modified amylose having
primary amino groups at some 60% of the primary
positions. Following strategy developed in the
monosaccharide field, the primary azide was
photolyzed with soft ultraviolet radiation in a 2-
methoxyethanol—benzene mixture to give 6-aldehydo
amylose, whose degree of substitution by aldehyde was
the same as that of the starting sulfonate and
intermediate azide; this extent of substitution was
independently verified by conversion of the product
into its p-nitrophenylhydrazone[3].

Similar procedures were used[5] with cellulose
derivatives in homogeneous media for introduction and
transformation of the azido group at C-6. However,
difficulties in achieving complete selectivity at
the primary position by a sulfonyloxy leaving-group[5]
led to the exploration of an alternative method[6] for
placing a leaving group at the primary position. As
illustrated (Fig. 3) for cellulose, heterogeneous
reaction of the polysaccharide with methanesulfonyl
chloride in N,N-dimethylformamide leads to the direct
formation of the 6-chloro derivative, and there is no
evidence for introduction of chlorine at the
secondary positions[7]. The degree of substitution may
be controlled within a wide range by appropriate
adjustment of the reaction conditions, permitting
access to 6-chlorocelluloses of d.s. from 0.1 to
0.8. The structure of the product was confirmed by
classical means through acid hydrolysis and isolation

of the crystalline α-tetraacetate of 6-chloro-6-deoxy-D-glucose.

The reaction works equally effectively for several different morphological forms of cellulose, in each instance under heterogeneous conditions. Introduction of primary chloro groups in the cellulose leads to marked changes in the properties of the polysaccharide, and at moderate d.s., the product becomes freely soluble in water. At higher d.s., it becomes once more water-insoluble.

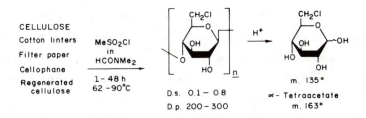

CELLULOSE
Cotton linters
Filter paper
Cellophane
Regenerated cellulose

MeSO₂Cl in HCONMe₂

1-48 h
62-90°C

D.s. 0.1-0.8
D.p. 200-300

H⁺

m. 135°

α-Tetraacetate
m. 163°

FIGURE 3   <u>6-Chlorination of Cellulose</u>

Primary chloride is readily displaced from 6-chloro-6-deoxycellulose by reaction with sodium azide in an appropriate dipolar aprotic solvent, to yield the corresponding 6-azido-6-deoxycellulose having negligible chlorine content and a d.s. by azide of

the same level as the original chloride (Fig. 4).
Little depolymerization at this stage is encountered.

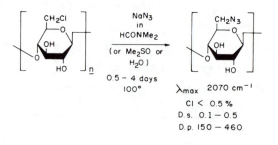

FIGURE 4   6-Azido-6-deoxycellulose

   Photolysis of the 6-azido-6-deoxycellulose
obtained by way of the 6-chloro derivative leads to
6-aldehydocellulose of the same d.s. as the original
chloride (Fig. 5).   The d.s. in the aldehydo product
may be readily estimated[3,5] by reduction with
deuterated sodium borohydride, followed by hydrolysis
and acetonation of the resultant glucose to form the
6-deuterated derivative of the well-known 1,2;5,6-di-
O-isopropylidene-α-D-glucofuranose. Mass-
spectrometric examination of this product and study
of the fragment-ion that incorporates C-5 and C-6
allows a measure, from the deuterium incorporation at
C-6, of the extent of substitution by aldehyde
groups; this is found to be the same as that of the
original 6-chlorocellulose.

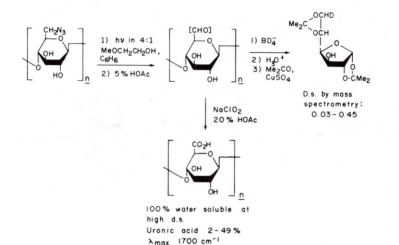

FIGURE 5   6-Aldehydocellulose

In addition to furnishing a very effective
procedure for preparing 6-aldehydocellulose (which
constitutes a model compound for study of the
alkaline oxidative behavior of cellulose), the route
also offers access, by oxidation with sodium
chlorite, to the corresponding 6-carboxycellulose,
obtained as a fully water-soluble polymer at
moderate-to-high levels of conversion. This product
may be regarded as cellulose in which controlled
proportions of glucuronic acid residues have been
incorporated by this chemical procedure.

Application of the same chlorination method to
pyridine-swollen amylose likewise[8] leads readily to
6-chloro-6-deoxyamylose (Fig. 6). At degrees of

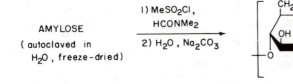

FIGURE 6   6-Chloro-6-deoxyamylose

substitution near 0.5, the product is a water-soluble,
white product, and at levels of substitution of 0.8 or
higher the product becomes water-insoluble.

The 6-chloro-6-deoxyamylose is readily converted
into the 6-azido analog by sodium azide in water, and
the product contains no residual chlorine (Fig. 7).
Photolysis of the azide with soft (300 nm)
ultraviolet radiation leads to 6-aldehydoamylose of
the same d.s. as the starting chloride; this product
shows the expected high lability to alkali and
undergoes rapid depolymerization by beta elimination
under alkaline conditions.

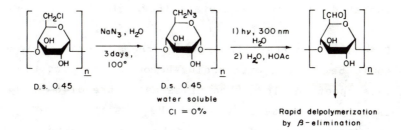

FIGURE 7   Photolysis of Azido Amylose

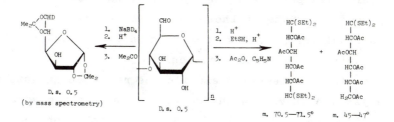

FIGURE 8   Characterization of 6-Aldehydoamylose

The constitution of 6-aldehydoamylose of d.s. 0.5 was established[3] by fragmentation methods.

Mercaptolysis yielded both the diethyl dithioacetal
of D-glucose, isolated as the pentaacetate, plus the
tetraethyl bis(dithioacetal) of D-_gluco_-hexodialdose,
isolated as its tetraacetate (Fig. 8).

As an alternative to this classical
fragmentation characterization, the aldehydo amylose
could be reduced with sodium borodeuteride and the
mass-spectrometic method of assay of the di-_O_-
isopropylidene derivative of the constituent
monosaccharide used[3,5,8] to determine the extent and
position of oxidation, as measured by the
incorporation of deuterium at the primary position.

3.2. Partial Acetylation Reactions

The literature records numerous examples of reactions
involving partial substitution of polysaccharides in
which specificity or selectivity for certain
positions, especially O-6 and O-2, has been
proposed[9]. Some of these claims have been
contradictory and few have been substantiated by
unequivocal characterization.

In an attempt to examine the relative extents of
acylation of various hydroxyl groups in carbohydrate
derivatives, model studies were undertaken on the
partial acetylation of methyl α-D-glucopyranoside[10].
The glucoside was allowed to react in pyridine
solution with 1 molar equivalent of acetylation
reagent to cause substitution to the net extent of 1
acetyl group distributed between all of the four
hydroxyl groups. Under the conditions used, it had

been clarified that no subsequent migration of acetyl
groups occurred. At this point, the acetylation was
taken to completion with deuterated acetic anhydride.
The resultant, fully acetylated methyl α-D-gluco-
pyranoside tetraacetate was isolated crystalline, and
the extent of incorporation of protioacetyl groups at
the various positions[11] determined by proton n.m.r.
spectroscopy.

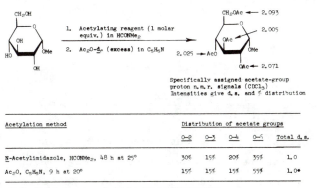

Specifically assigned acetate-group
proton n.m.r. signals (CDCl$_3$)
Intensities give d.s. and % distribution

Acetylation method	Distribution of acetate groups				
	O-2	O-3	O-4	O-6	Total d.s.
N-Acetylimidazole, HCONMe$_2$, 48 h at 25°	30%	15%	20%	35%	1.0
Ac$_2$O, C$_5$H$_5$N, 9 h at 20°	15%	15%	15%	55%	1.0*

*Similar distribution observed with β anomer but slower reaction.

FIGURE 9   Acetylation of Model Glucoside

The results (Fig. 9) demonstrate[10] that there is
a higher extent of acetylation at the primary
position, but there is little difference in extent of
acetylation at the three secondary positions, even
when N-acetylimidazole, a reagent supposedly
selective for primary positions, was used.

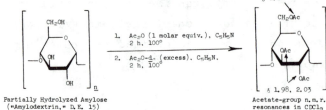

Partially Hydrolyzed Amylose
("Amylodextrin," D.E. 15)

1. Ac₂O (1 molar equiv.), C₅H₅N
    2 h. 100°
2. Ac₂O-d₃ (excess). C₅H₅N.
    2 h. 100°

Acetate-group n. m. r.
resonances in CDCl₃

Distribution of acetate groups

0-2 and 0-3, 60%; 0-6, 40%; total d.s. 1.0

FIGURE 10   Partial Acetylation of Amylose

Application of the same procedure at the
polysaccharide level with amylose[12] (Fig. 10)
likewise showed that acetylation was more extensive
at the primary position, but that significant
acetylation occurred at both secondary positions;
there was little difference in the extent of
acetylation at O-2 and O-3.

A very promising method[13] for selective
substitution at the primary position of such
polysaccharides as amylose and cellulose involves
initial preparation of the per(trimethylsilyl) ether,
followed by treatment of this product in carbon
tetrachloride with pyridine, acetic anhydride, and
acetic acid under controlled conditions. Exclusive
replacement of the primary substituent by acetoxyl
takes place.   The resultant product can be readily

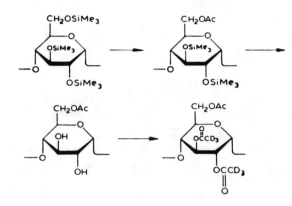

FIGURE 11   Selective 6-0-Acetylation of Amylose

desilylated at the secondary positions to afford, for
instance, amylose specifically acetylated at O-6
(Fig. 11).

Again an n.m.r.-spectroscopic method was used to
identify the position of acylation in this product,
and it was established that no detectable
substitution at the secondary hydroxyl positions
takes place.   This reaction holds high promise as a
more-versatile method for protection of glycans at
the primary position than the traditional method of
tritylation.

4.   UNSATURATED AND DEOXYGENATED POLYSACCHARIDES

4.1.   5,6-Unsaturation

The specific introduction of points of
unsaturation into a polysaccharide backbone
constituted an important part of our program.   Such a

D. HORTON

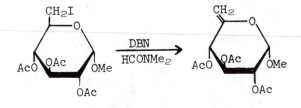

DBN = 1,5-diazabicyclo[4.3.0]non-5-ene

FIGURE 12   Model System for 5,6-Elimination

reaction was evaluated in a model monosaccharide
system (Fig. 12) in which the 6-deoxy-6-iodo
derivative of methyl α-D-glucopyranoside, as its
triacetate, was subjected to base-catalyzed
elimination[14]. The reaction gives a 5,6-unsaturated
sugar, a product that is formally a cyclic enol
ether.

This same reaction may be effected at the
polysaccharide level, as demonstrated (Fig. 13) with
6-deoxy-6-iodoamylose prepared by way of the
corresponding 6-O-tosyl derivative[15]. The
elimination reaction likewise gives the unsaturated
sugar as the point of modified functionality in the
polysaccharide chain, with the same extent of
substitution as the original iodide.

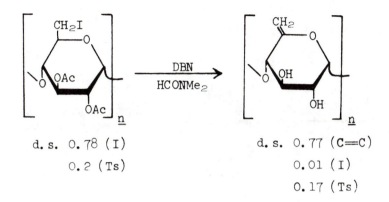

d. s. 0.78 (I)
     0.2 (Ts)

d. s. 0.77 (C=C)
     0.01 (I)
     0.17 (Ts)

FIGURE 13   5,6-Unsaturated Amylose Derivative

The constitution of this product was verified degradatively by acid hydrolyis; the predicted high acid lability of the enol ether leads to liberation of the constituent monosaccharide, characterized as the bis(p-nitrophenylhydrazone) of 6-deoxy-D-xylo-hexos-5-ulose (Fig. 14).

This sequence of reactions verifies the enol ether structure of the starting material and demonstrates that the iodide elimination had not proceded by the alternative, possible reaction, namely, formation of 3,6-anhydro linkages in the polysaccharide chain. Were this reaction to be reinvestigated, the starting material of choice would be the 6-chloro-6-deoxy rather than the 6-O-tosyl derivative, for reasons earlier mentioned.

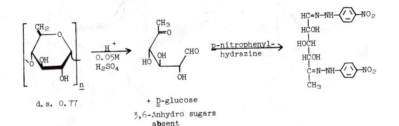

d.s. 0.77

\+ D-glucose

3,6-Anhydro sugars
absent

FIGURE 14   Degradation of 5,6-Unsaturated
Amylose Derivative

4.2.   6-Deoxygenation

6-Deoxy-6-iodoamylose also offers access to a
deoxygenated polysaccharide, 6-deoxyamylose, by
photolysis of the iodide in methanol solution in the
presence of an acid acceptor[15] (Fig. 15).

The iodide is redily cleaved, presumably by a
free-radical process, to furnish the corresponding 6-
deoxy polysaccharide, which was demonstrated to give
6-deoxy-D-glucose on hydrolysis.

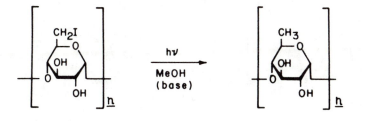

FIGURE 15   6-Deoxyamylose

4.3.   2,3-Unsaturation

The introduction of double bonds into the
polysaccharide matrix other than between C-5 and C-6
is more difficult, but may be accomplished at the
2,3-position[16,17] by treatment of an appropriate 2,3-
disulfonate with sodium iodide in N,N-
dimethylformamide in the presence of an excess of
zinc dust.   In the case of amylose, the primary
position must initially be protected as by the 6-O-
trityl derivative.   The 2,3-di-p-toluenesulfonate
reacts under the specified conditions to give the
2,3-unsaturated derivative.   Traces of 3,6-anhydro-D-
glucose residues detected must have arisen by partial
detritylation and primary tosylation during the
tosylation step (Fig. 16).

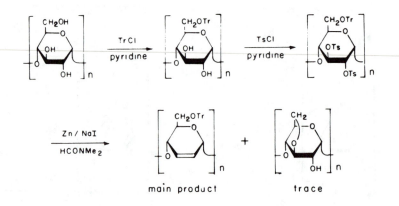

FIGURE 16　2,3-Unsaturated 6-0-Tritylamylose

The resultant unsaturated polysaccharide is very
acid-labile, as expected because the anomeric
position is now allylic. The compound undergoes
depolymerization in aqueous acetic acid to give
triphenylmethanol plus 2-(D-glycero-1,2-
dihydroxyethyl)furan, characterized as its
crystalline bis(p-nitrobenzoate), a compound prepared
as a reference standard by a known sequence of
reactions from a corresponding monosaccharide
derivative containing 2,3-unsaturation (Fig. 17).

This unsaturated amylose derivative shows the
expected behavior upon reduction with hydrogen or
deuterium, and the reduction is conveniently
monitored by n.m.r. spectroscopy. Addition of
bromine takes place readily to give a compound whose
elemental analysis corresponds to the anticipated
dibromide structure[16,17] (Fig. 18).

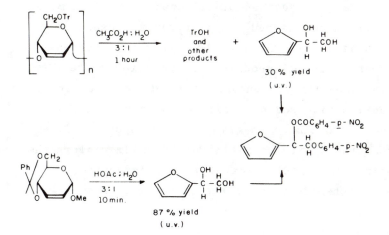

FIGURE 17   Degradation of 2,3-Unsaturated Amylose

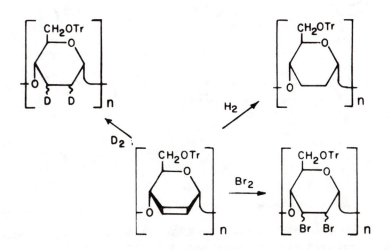

FIGURE 18   Reactions of 2,3-Unsaturated Amylose

The introduction of 2,3-unsaturation is more
readily accomplished in polymeric pentoses, as
exemplified by xylan, because this structure has no
primary hydroxyl group requiring prior protection.
Thus, the di(p-toluenesulfate) of xylan undergoes
reaction with sodium iodide—zinc dust in N,N-
dimethylformamide to give the corresponding 2,3-
unsaturated polymer fully functionalized along the
chain[17] (Fig. 19). Again, this unsaturated
polysaccharide is very labile to acid, and is
converted by aqueous acetic acid into 2-
(hydroxymethyl)furan.

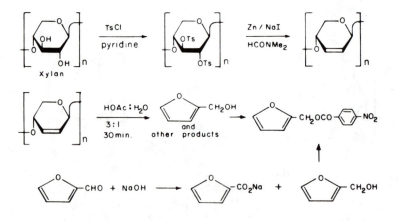

FIGURE 19   2,3-Unsaturated Xylan

5. SPECIFIC OXIDATION OF POLYSACCHARIDES

5.1. Primary Position

The indirect oxidation of primary hydroxymethyl
groups in polysaccharides to aldehydes by photolysis
of azide intermediates has already been presented.
Reduction of such azides likewise provides a method
for introduction of primary amino functionality into
the sugar residues of polysaccharides. The specific
introduction of secondary amino functionality into
polysaccharides has been a concurrent goal in our
laboratory for a number of years. Efforts to conduct
nucleophilic displacement-reactions at secondary
positions as a method for introducing nitrogen
functionality have met with limited success at the
polysaccharide level[18]. However, the ease of
introduction of amino functionality into
monosaccharides by reduction of oxime intermediates
led to a sustained effort to accomplish specific
oxidation at secondary positions in polysaccharide
derivatives, to provide precursors for oximes and
thence of the aminopolysaccharide targets.

5.2. Secondary Positions: Aminated Polysaccharides

Amylose protected by tritylation at the primary
position undergoes selective oxidation at one of the
two remaining hydroxyl groups by the reagent dimethyl
sulfoxide—acetic anhydride, with high selectivity
for the 2-position. The resultant 2-keto derivative

may be oximated and subsequently reduced with lithium
aluminum hydride in tetrahydrofuran to give an
aminated amylose bearing most of the amino
substitution at position 2 in the D-glucose
stereochemistry[19]. Degrees of substitution up to 0.7
may be achieved (Fig. 20). A troublesome feature was
incorporation of some (methylthio)methyl ether groups
at O-3.

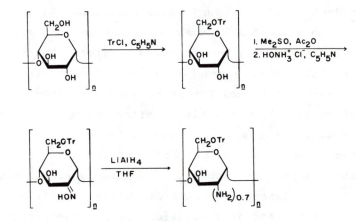

FIGURE 20   C-2 Oxidation of 6-O-Tritylamylose

Trifluoroacetylation of this product, followed
by detritylation (Fig. 21) yields a protected
aminated amylose derivative having the primary
position unsubstituted.

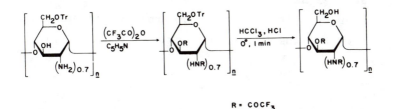

FIGURE 21   Protected 2-Amino-2-deoxyamylose

Efforts to oxidize the primary positions by the azide photolysis route met with limited success, but oxidation with atmospheric oxygen in the presence of platinum dioxide led to partial oxidation of the primary positions, to d.s. 0.25   (Fig. 22).

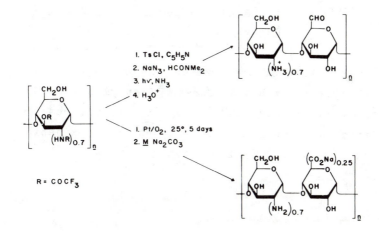

FIGURE 22   C-6 Oxidized 2-Amino-2-deoxyamylose

This material, from which the protecting groups
at N-2 and O-3 had been removed, was sulfated to
afford a product having functionality emulating that
of heparin, and which displayed some anticoagulant
activity (Fig. 23).

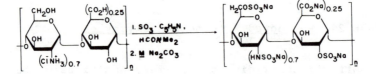

FIGURE 23   Sulfated 2-Amino-2-deoxyamylose
             Derivative

Evaluation of the same precursor (6-O-
tritylamylose) with a different dimethyl sulfoxide-
based oxidation system[20] led again[21] to favored
oxidation at C-2; a smaller proportion of oxidation
occurred at C-3 (Fig. 24).

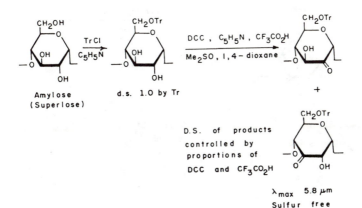

FIGURE 24   <u>Oxidation of 6-0-Tritylamylose</u>

The extent of reaction at each position was
assayed for total extent of oxidation by conversion
into the oxime and analysis for nitrogen, and also by
analysis of the products after borohydride reduction,
detritylation, and subsequent spectroscopic analysis.
Concurrent analyses involved hydrolysis of the
reduced polysaccharide derivative and subsequent
quantitation of sugars in the hydrolyzate (Fig. 25).

The results (Table I) show that the degree of
oxidation may be controlled by the proportions of
reagents used and the reaction conditions, and range
from very low d.s. to material having d.s. 0.7.

D. HORTON

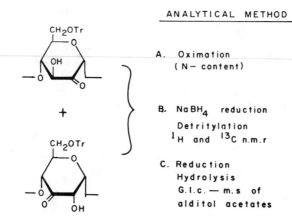

ANALYTICAL METHOD

A. Oximation
   (N– content)

B. NaBH₄ reduction
   Detritylation
   ¹H and ¹³C n.m.r

C. Reduction
   Hydrolysis
   G.l.c. — m.s of
   alditol acetates

FIGURE 25   Analysis of Oxidized 6-O-
             Tritylamylose

Analysis of DCC—CF₃CO₂H-Oxidized 6-O-Tritylamylose

Expt. No.	DCC/CF₃CO₂H per mol.	D.S. by oximation	Results of g.l.c. analysis[a]			Mol. wt.[b]	Iodine
			Glc	Man	All		
1[c]	5:1	70%	75%	4%	21%	8,000	–
2[d]	4:1	80%	93%	3%	4%	~20,000	–
3[d]	2:0.5	35%	95%	1%	4%	~20,000	+
4[d]	1:0.25	20%	9%	trace	trace	~35,000	+

[a] Supported by ¹³C and ¹H n.m.r. analysis of reduced, detritylated product.

[b] By gel-permeation chromatography of reduced, detritylated product.

[c] Superlose 149974.

[d] Superlose 394554.

TABLE I   Analytical Data for Oxidized Amyloses

The sugars recovered demonstrate that oxidation
was very specific at C-2 at low degrees of oxidation,
as no allose was detected in the hydrolyzate of the
reduced polysaccharide.   As the degree of oxidation
increased, selectivity between the two secondary
hydroxyl positions decreased.   The onset of oxidation
at C-3 is signaled by the significant proportion of
allose residues in the reduced polysaccharide.

6-$\underline{O}$-Tritylamylose oxidized by this procedure to
degree of substitution 0.35 was oximated to give a
sulfur-free product, in contrast to the product of
the earlier oxidation procedure, which retained some
sulfur arising from incorporation of
(methylthio)methyl ether groups (Fig. 26).

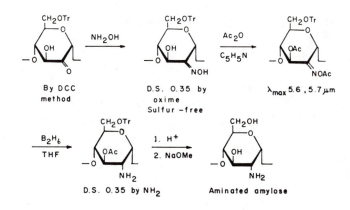

FIGURE 26   **Preparation of Aminated Amylose**

An improvement in the procedure for conversion
into the amine was realized by reducing the
acetylated oxime with diborane. Deprotection of the
reduced product led to an aminated amylose of the
same d.s. as the original oxidized product. The
sugar residues in this material are shown to be 2-
amino-2-deoxy-D-glucose, which was identified in the
hydrolyzate of the polysaccharide. The properties of
the product (Fig. 27) are concordant with the
structure of an α-(1$\rightarrow$4)-linked glycan containing 2-
amino-2-deoxy-D-glucose residues, together with
unmodified D-glucose residues.

Characterization and Properties of
Aminated Amylose

Properties		Analysis	
Ninhydrin positive		(alditol acetates	
Iodine blue complex		from hydrolyzate)	
Soluble in : H_2O , Me_2SO		Glc	74 %
dilute acid base		GlcN	23 %
I.r, ^{1}H n.m.r no NHAc,		ManN	3 %
δ 5.34 (H−1 of GlcN),		AllN-3	0%
5.25 (H−1 of GlcN)		GlcN-3	0%

Aminated amylose
(D.S. 0.35)
by DCC route

Mol. wt. : 4,000 − 20,000
(Sephadex G−25, G−75)
ID_{50} 7×10^{-6}M (L−1210 cells)

FIGURE 27   Characteristics of Aminated Amylose

The negligible content of other amino sugars in
hydrolyzates of this product demonstrates that
involvement at C-3 is minimal at this level of
oxidation.

The structural attribution of this product is
further substantiated by independent physical
studies, as by high-field (125 MHz) carbon-13 n.m.r.
spectroscopy[22] (Fig. 28), which shows a pattern of
carbon resonances resembling that of the parent
amylose but with additional carbon resonances near 60
p.p.m. characteristic of carbon atoms bearing
nitrogen substituents.

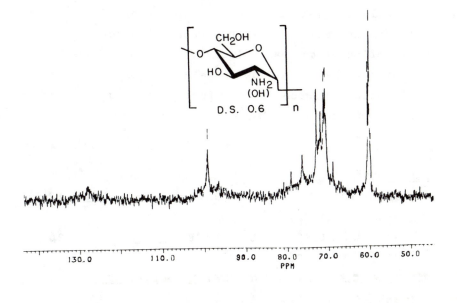

FIGURE 28 ^{13}C-N.M.R. Spectrum at 125 MHz of 2-
Amino-2-deoxyamylose

6. CONCLUSIONS

The foregoing results demonstrate very broadly that
many of the transformations effective with
monosaccharides and their derivatives may also be
accomplished at the polysaccharide level, given due
attention to the different constraints of operating
within the polymer matrix. In general those
reactions, feasible under mild conditions with
monosaccharide derivatives and which can be adapted
for reaction with polysaccharides in homogeneous or
heterogeneous media, have been the most successful.
With the newer chemical and physical methods
available for characterization of polysaccharides,
especially high-field n.m.r. spectroscopy of
polysaccharides, there is now much scope for
utilization of the basic leads uncovered by these
investigations for a wide range of useful
applications of specifically functionalized
polysaccharides of defined properties.

REFERENCES

1. D. HORTON and W. D. PARDOE, Carbohydr. Res., **12**,
    269—272 (1970).

2. R. L. WHISTLER AND S. HIRASE, J. Org. Chem., **26**,
    4600—4605 (1961).

3. D. M. CLODE AND D. HORTON, Carbohydr. Res., **17**,
    365—373 (1971).

4. C. J. MALM, L. J. TANGHE, AND B. C. LAIRD, J. Am.
    Chem. Soc., **70**, 2740—2747 (1948).

5. D. M. CLODE AND D. HORTON, Carbohydr. Res.,
   19, 329—337 (1971).

6. M. E. EVANS, L. LONG, JR., AND F. W. PARRISH, J.
   Org. Chem., 33, 1074—1076 (1968).

7. D. HORTON, A. E. LUETZOW, AND O. THEANDER,
   Carbohydr. Res., 26, 1—19 (1973).

8. D. HORTON, A. E. LUETZOW, AND O. THEANDER,
   Carbohydr. Res., 27, 268—272 (1973).

9. J. M. SUGIHARA, Adv. Carbohydr. Chem., 8, 1—44
   (1953).

10. J. GELAS AND D. HORTON, Carbohydr. Res.,
    64, 323—326 (1978).

11. D. HORTON AND J. H. LAUTERBACH, J. Org. Chem.,
    34, 86—92 (1969).

12. D. HORTON AND J. H. LAUTERBACH, unpublished data.

13. D. HORTON AND J. LEHMANN, Carbohydr. Res., 61,
    553—556 (1978).

14. D. HORTON AND R. G. NICKOL, unpublished data.

15. E. S. H. EL ASHRY AND D. HORTON, unpublished data.

16. D. M. CLODE, D. HORTON, M. H. MESHREKI, AND H.
    SHOJI, Chem. Commun., 694—695 (1969).

17. D. HORTON AND M. H. MESHREKI, Carbohydr. Res.,
    40, 345—352 (1975).

18. M. L. WOLFROM, M. I. TAHA, AND D. HORTON, J. Org.
    Chem., 28, 3553—3554 (1963).

19. D. HORTON AND E. K. JUST, Carbohydr. Res.,
    30, 349—357 (1973).

20. A. I. USOV AND V. S. IVANOVA, Izv. Akad. Nauk
    SSSR, Ser. Khim., 4, 910—915 (1973).

21. D. HORTON AND T. USUI, Carbohydrate Sulfates, ACS
    Symp. Ser., 77, 95—112 (1978)

22. C. W. BERGMANN AND D. HORTON, unpublished data.

NEW DERIVATIVES OF CHITIN AND CHITOSAN:

PROPERTIES AND APPLICATIONS

R.A.A. MUZZARELLI
Faculty of Medicine, University of Ancona,
I-60100 Ancona, Italy

After discussing the treatments of chitinous materials and the specifications of chitin and chitosan
to be used industrially, this Chapter presents the
current trends in several application fields, with
emphasis on the uses of chitosan membranes in medicine and in biotechnology, the chemical derivatives
of chitosan, especially those obtained by Schiff
reaction and N-acylation and the chelation of metal
ions by a number of chitosan derivatives.

1. INTRODUCTION

Most native chitinous and cellulosic tissues are composites
in which the fibers are suspended in a lower modulus matrix.
In chitinous tissues, the matrix is usually protein, and may
also be calcified or sclerotized; most of these tissues are
structural complexes of chitin and protein in which both pha
ses are ordered. In plant cell walls, the cellulose fibrils
are in a matrix of amorphous hemicelluloses (xylans and man-
nans) accompanied, in wood, by amorphous cross-linked lignin.
    The remarkable fact that chitin does not occur in higher
plants where cellulose is the major polysaccharide would be

sufficient to sustain the view that chitin is not a cellulo-
se-like polysaccharide, however, further aspects of diversity
can be pointed out. Many attempts to assemble cellulose microfibrils in vitro have been made without success, yet micro
fibrils of chitin and chitosan have been recently synthesized
in vitro, taking advantage of insect and fungal preparations.
The high nitrogen content of chitin/chitosan as opposed to
the absence of nitrogen in cellulose confers to chitin/chito
san the characteristic properties of cationic polysaccharide
because the nitrogen atoms constitute amino groups which are
acetylated to various extents. Of course, the ß-D-(1-4) glucan chain is a feature common to both chitin and cellulose,
however this point of similarity in the primary structure is
compensated by the different hydrogen bonds involving the
acetamido groups of chitin and leading to the filmogenic pro
perties of chitin and chitosan which are not shared by cellu
lose. While most of the chemistry of cellulose is common to
chitosan, all of the chemical reactions involving the primary
amino groups of chitosan are of course unfeasible on cellulo
se: they include important reactions such as the protonation
of the amine which enables the dissolution of chitosan by a
large number of strong and weak acids.

    If some of the above differences are the reason of the
currently scarce use of chitin, on the other hand the peculiar properties of chitin/chitosan attract more and more
scientists seeking the possibility of exploiting chitin for
the production of high added value items in various scientific and technological fields [1 - 15].

2. PROCESSING OF CHITINOUS MATERIALS

Chitin is easily isolated from crab or shrimp shells or from
fungal mycelia. In the first case, the production of chitin
is associated with the fishing activities, the animals treated being mainly the shrimps Pandalus borealis and Euphausia
superba and the crabs Chionoecetes opilio, japonicus and
bardii; in the second case, the production of chitosan-glucan
complexes is associated with fermentation processes, such as
that for the production of citric acid with Aspergillus niger [16]. The canning industries are mainly based in Virginia,
Oregon, Washington, and in Japan and receive supplies from

fishing fleets in the Antarctic Sea. Several countries   have
large unexploited crustacean resources, for instance Norway,
Mexico and Chile [2, 6].

The processing of crustacean shells mainly involves the
removal of proteins and the dissolution of calcium carbonate;
the so obtained chitin is then submitted to deacetylation in
40 % sodium hydroxide at 120°C for 1-3 h to obtain chitosan,
with the deacetylation degree of about 70 %. Fully deacetylated chitosan can be obtained by repeating the alkali   step.

Two general trends in the industrial production   and
uses of chitin/chitosan can be noticed when the production
data and the value of the products are analyzed. The   first
trend is the increase of chitin production for industrial
needs; for instance, in the case of one major Japanese pro-
ducer, the increase over the 5-year period was 37 % as shown
in Table I. The major portion of chitosan goes into   sludge
dewatering (25 % increase) and smaller portions find   their
uses in food processing (300 % increase) and in metal   ion
chelation (100 % increase). Table I does not include   small
amounts of chitosan used to produce intermediate and   high
value items.

TABLE I   Data on chitosan production from Tanner   crab   and
Scarlet Queen crab (Chionoecetes opilio and japanicus)   and
their major uses. Values in tons per year. Courtesy of Kyowa
Yushi Kogyo Co., Japan.

Year	Sludge dewatering	Food processing	Metal chelation	Membranes	Total
1978	205	8	10	4	227
1979	203	14	13	2	232
1980	158	15	15	2	190
1981	147	24	18	3	192
1982	204	30	20	4	258
1983	255	34	20	2	311

The second trend concerns the quality of the   chitosan
and the value of the products: the use of chitosan   evolves
today towards intermediate value items such as cosmetics,
drug carriers, animal feed additives, membranes and   so   on,

while high value chitin-based products will probably be developed soon mainly in the field of nutrition, immunology,
medical aids and pharmaceuticals. The pharmaceutical industry
currently makes use of significant quantities of glucosamine
which is obtained from chitin.

Only when the crustacean shells become available as a
waste from a food industry, the production of chitosan   is
economically feasible; preferably the recovery of carotenoid
is included in the process. The shells contain   considerable
quantities of astaxanthin, a carotenoid whose chemical synthesis has so far been unsuccessful and which is marketed
as a fish feed additive in aquaculture.

3. SPECIFICATIONS FOR CHITIN/CHITOSAN

Depending on which uses are anticipated, the   specification
sheet of chitosan has to include a number of data   such   as
those listed in Table II. The degree of acetylation   is   an
important parameter for the chemical characterization and is
determined by gas-chromatography and by infrared   spectrometry [17]. Of particular interest because of its simplicity and
precision is the determination of the degree of acetylation
by first derivative ultraviolet spectrophotometry, which is
especially suited for highly deacetylated chitosans. By measuring absorption at 199 nm in the first derivative   mode,
both the spectroscopic contributions of acetic acid (needed
to dissolve chitosan) and of accompanying   glucosamine   are
offset, because the wavelength corresponds   to   the   acetic
acid zero crossing point, and because the 199 nm peak for
N-acetylglucosamine is 130 times higher than the glucosamine
contribution [18]. Of course, the moisture content of the chitosan submitted to titration as well as the pH value of the
medium from which it was isolated should be   taken   into
account for accurate measurements. Table III shows the data
for six commercial chitosans and two experimental chitosans,
obtained by first derivative UV spectrophotometric   method;
the results are more precise than those obtained by IR spectrometry on the same samples. Spectra are shown in Figure 1.

In certain cases, for instance when studying   the   differences between chitins of diverse origins or when dealing
with   modified   chitins, [13]C n.m.r.   and mass spectrometry

TABLE II   Specifications for chitin and chitosan.

Nitrogen, usually from 6 to 7 % in chitins, from 7 to 8.4 % in chitosans.

Degree of acetylation, usually around 90 % in chitins, 40 % in current chitosans, between 10 and 0 % in fully deacetylated chitosans.

Viscosity of 1 % chitosan solutions in 1 % acetic acid, from 200 to 3000 cps; degraded chitosans have lower values.

Molecular weight, native chitins have m.w. in excess of one million dalton; commercial chitins and chitosans have m.w. in the range 1 to 5 x 10^5 dalton. Polydispersity is also an important parameter.

Titrations, the moisture content and the pH of the water from which chitosans were isolated should be taken into account.

Dissociation constant, K_a is between 6.0 and 7.0, most often 6.3.

X-ray diffraction data, typical peaks at 8°58' to 10°26' and at 19°58' to 20°00'.

Moisture, from 2 to 10 % under laboratory conditions.

Impurities:

Ashes at 900°C: usually lower than 1 %.

Carotenoids, it should be indicated whether they have bee extracted or bleached.

Aminoacids, traces of glycine, serine and aspartic acid may be present.

Transition metals, with the exception of iron, normally below a total of 5.0 µg/g.

---

can provide information. Pyrolysis-mass spectra of chitosan samples of different degrees of acetylation indicated that as the degree of acetylation decreases, the peak ratios 80/60, 67/60 and 80/42 increase to a limit representing the limit of deacetylation of chitin. The 80 and 67 fragments originate from the D-glucosamine moiety of the polymer and the 60 and 42 fragments from the N-acetylglucosamine moiety. The abundant fragment at m/z 59 was identified as acetamido and in fact it is less pronounced in the chitosan spectrum [19].

TABLE III   Degrees of acetylation of several commercial chitosans analyzed by first derivative ultraviolet spectrophotometry. Muzzarelli, original results.

Chitosan	Mean of five measurem.	Confidence limits at 95 % level	Relative standard dev. %
Bioshell	27.5	0.8	2.2
Kyowa	24.5	0.8	2.8
Chesapeake	27.9	0.6	1.9
Anic	26.3	0.8	2.5
Polyplate	23.7	1.0	3.5
Rybex	42.6	0.7	1.3
Rybex, deacetylated	4.5	0.3	6.0
Rybex, sonicated	18.4	0.8	4.2

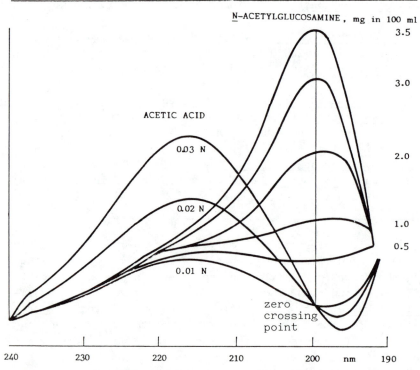

FIGURE 1.   First derivative ultraviolet spectra for acetic acid and for N-acetylglucosamine at various concentrations.

The description of a ^{13}C n.m.r. spectrum for a chemical
derivative of chitin is given later. It should be remarked
that in certain cases profound alterations are introduced in
the typical spectrum; for instance, the spectrum obtained on
chitin dissolved in a saturated solution of lithium thio-
cyanate shows the C-3 and C-5 signals superimposed at δ =
74.8, whereas no signal of the N-acetyl group is visible,
neither the CH_3 in the region δ = 20, nor the carbonyl in the
region δ = 170, due to strong interactions of the N-acetyl
group with the lithium ion [20].

Of course, not every chitosan preparation should comply
with stringent specifications, but, if a chitosan batch is
to be used in photosensitive materials or in medical aids,
for instance, its impurities should be declared and kept as
low as possible.

4. CONFORMATION OF CHITOSAN IN SOLUTION

Because most of the uses of chitin are made possible by the
preparation of chitosan solutions, the rheological behavior
of chitosan, a typical random coil polyelectrolyte, is an
important subject of study. Only recently, some analysis of
the solution behavior of chitosan has begun [3, 21 - 23].

The flow behavior of polyelectrolyte polylsaccharides
is dictated by their overall molecular conformation and the
degree of hydrogen bonding or electrostatic repulsion between
neighboring chain segments. The intrinsic viscosity determined for chitosan in solution indicates that chitosan adopts
conformations which range from a random coil to a more compact "quasi-globular" shape. The intrinsic viscosity, $[\eta]$,
dl/g, of a linear chain polymer follows a dependence on molecular weight according to the Mark-Houwink equation:

$$[\eta] = k.M^a$$

In the case of random coil polymers, the exponent "a" usual
ly lies between 0.5 and 0.8, and the overall shape of the
hydrodynamic volume is spherical. Exponents greater than 0.8
reflect a highly expanded conformation which may be rod-like
in shape. Low values indicate globular conformation that can
be caused by local ordering due either to attractive hydrogen
bonding forces or to electrostatic repulsive forces. Studies
on the relationship between intrinsic viscosity of chitosan

and the molecular weight demonstrated the significance of
hydrogen bonding on the molecular conformation of chitosan
in solution. This is shown in Table IV; these different "a"
values represent the conformational changes in chitosan from
a compact hydrodynamic sphere to an extended random coil,
when hydrogen bonding is suppressed by urea. For comparison,
alginate and hyaluronate have Mark-Houwink exponents approaching 1.0 because they have a high degree of chain extension
in solution.

TABLE IV   Mark-Houwink exponent for chitosan from Chionoecetes bairdi (Tanner crab) with degree of deacetylation 80 %
and molecular weight 130,000.

Solvent	Exponent a
Trifluoroacetic acid	0.296
Acetic acid 1 %, sodium chloride 2.8 %	0.147
Acetic acid 1 %, lithium chloride 2.0 %	0.186
Acetic acid 0.2 M, sodium chloride 0.1 M, urea 4 M	0.710

As for all the polyelectrolyte polysaccharides, the
Mark-Houwink exponent is pH- and ionic-strength-dependent:
therefore, the hydrodynamic volume of chitosan can be mani-
pulated by altering the intramolecular interactions between
charged neighboring ions by simple changes in pH and ionic
strength.

5. CHITOSAN MEMBRANES

The film-forming ability of chitosan has been the object of
many studies, some of which have led to industrial uses in
various fields. Photographic films, reverse osmosis membranes and cosmetics containing significant amounts of chitosan
are now being marketed.

Procedures for preparing photographic images by diffusion transfer include chitosan as a protective layer, together with copper(II) salts and glycerol [24] or gelatin [25].
Relatively thick protective layers can be used to enhance
the long term stability and integrity of the films. The chelation of copper(II) ions by chitosan in form of membrane
has been studied in detail [26].

Reverse osmosis membranes were found to possess resistance to high alkali concentrations and being durable in some organic solvents [27]. Hollow chitosan fibers were spun with a special spinneret: the hollow fiber was coagulated externally with 1 N NaOH solution and internally with gaseous ammonia at dragging speed of 14 m min^{-1}. The fiber properties were: 17 g km^{-1}, diameter 0.5 mm, wall thickness 0.015 mm and rupture pressure > 600 mm Hg [28].

The use of chitosan as an ingredient of shampoos and lotions is also justified by the characteristic properties of chitosan membranes. A thin chitosan membrane having favorable gas permeabilities, forms on the hair and confers shine and strength; moreover, the hair treated with chitosan does not become statically charged during brushing [3, 29, 30].

Chitin too lends itself to membrane formation: chitin membrane can be cast from chitin solutions such as the mixture of formic acid and dichloroacetic acid, trichloroacetic acid and dichloroethane, and N,N-dimethyl acetamide, N-methyl 2-pyrrolidone and lithium chloride, the coagulating agents being 2-propanol or acetone [1].

Further applications of the film-forming ability of chitosan are in Section 7 of this Chapter.

5. CHEMICAL DERIVATIVES OF CHITIN AND CHITOSAN

In common with other polysaccharides, chitin undergoes O-substitution reactions to give esters and ethers whose physical properties have potential applications. Glycol chitin, a partially O-hydroxyethylated chitin, was probably the first derivative to find practical use and to be listed in catalogs of chemical products. For the determination of the enzymatic activity of lysozyme a method was developed [31] that uses as substrate glycol chitin synthesized using 2-chloroethanol instead of ethylene oxide. The relationship between the degree of substitution of chitin and the ability of glycol chitin to act as a substrate for lysozyme was investigated.

Many of the more recently prepared derivatives of chitin for which potential uses can be envisaged, are obtained according to two main approaches, each involving partial or complete removal of the native N-acetyl groups. The new substances are N-acyl derivatives obtained from acyl halides or

anhydrides, and the derivatives formed by hydrogenation of
the products of the Schiff reaction with aldehydes and ketones.

6.1. N-Acyl derivatives of chitosan

The N-acetylation of chitosan leads to fully N-acetylated
chitin. In studying the influence of the reaction medium on
the N-acetylation of chitosan, it was found [32] that maximum
reaction rate is achieved using binary mixtures of ethanol
and methanol or methanol and formamide. Complete N-acetylation may be achieved in three minutes at room temperature
using a highly swollen chitosan in organic aprotic solvents
[33]. Alternatively, the preparation of partially acylated chi
tosans was made by treating solutions of chitosan in acetic
acid-methanol with acid anhydrides at room temperature. Rapid N-acylation of chitosan can be achieved, in fact, under
heterogeneous conditions if highly swollen gels are prepared
or the sample has been steeped in solvent [33 - 35]. The struc
ture of the acyl residue has considerable influence on the
ease of O-acetylation, and with bulky acyl groups, steric
hindrance may prevent substitution beyond 0.5. Under these
conditions, the formation of highly acetylated products requires prolonged reaction times.

Long-chain di-O-acyl chitins may be prepared upon reaction with acyl chlorides or anhydrides in methanesulfonic
acid at low temperature [36]. Chitosans boiled with a large
excess of hexanoyl, decanoyl or dodecanoyl chlorides in dry
pyridine-chloroform afford fully acylated derivatives [37].
Highly benzoylated chitin shows good solubility in several
organic solvents [38]. An aspirin carrier was prepared by
reaction of chitosan with 2-acetoxybenzoic anhydride [39].

Carboxyacyl chitosans have been prepared from reactions
with anhydrides of alkyl and aryl dicarboxylic acids [40 - 42].
Partially succinylated chitosans and O-hydroxyethyl chitosan
containing both amino and carboxyl groups show varying solubility in water, dilute acid and dilute alkali, depending on
the degree of substitution [41]. By further reaction with car-
bodiimides, gels with a low degree of cross-linking have
been prepared [41, 43]. Some of these derivatives show solubi-
lities analogous to proteins, in being soluble in dilute so-

dium hydroxide, precipitating at pH 4 - 5 and dissolving   at low pH values.

Although the maleylation of amino groups   in   proteins may be carried out in water at pH 7.1 - 9.5,   the   heterogeneous reaction with chitosan was better performed in formamide [44]. The copolymers of N-maleylchitosan and acrylamide are stable at all pH values; they swell in   water   and   give gels with good mechanical properties.

6.2. Schiff bases of chitosan and their reduction products

The Schiff reaction between chitosan and aldehydes and ketones gives the corresponding aldimines and   ketimines,   which can be hydrogenated to products less susceptible   to   hydrolysis [45]. N-Alkyl chitosans from simple aldehydes   and   ketones have been isolated as white powders with degrees of substitution 23 - 33 %. The chelating ability was   specific for certain cations such as copper(II), mercury(II) and lead(II) for which the capacities were in the range 2 - 6 % by weight depending on conditions and choice of N-alkyl chitosan [46]. Intramolecular hydrogen bonds are apparently weakened by the presence of the bulky substituents and thus,   despite   their hydrophobicity, the N-alkyl chitosans swell   enormously   in water; they retain the film-forming ability of chitosan and membranes can be cast from their acetic acid solutions.

The reductive amination reaction of chitosan with polyfunctional aldehydes and ketones provide access to polyampho lytes. The addition of phthalaldehydic acid   or   glyoxylic acid to aqueous suspensions of chitosan resulted in immediate dissolution, with accompanying gel formation   at   suitable pH values [47 - 49]. The Schiff bases, upon hydrogenation, afforded N-(o-carboxybenzyl) chitosan (NCBC)   and   N-(carboxymethyl) chitosan (NCMC), soluble in both acidic and alkaline media (see Figure 2). In this compound, each   carboxymethylamino residue is simultaneously a secondary   amine   and   a carboxylic acid. These polyampholytes precipitate in the pre sence of transition metal ions because they   form   insoluble metal chelates: in the polysaccharide concentration interval studied (200 - 500 µg ml^{-1}), Co, Ni, Cu, Cd, Pb and U could be completely removed from 0.1 and 0.2 mM solutions by either NCMC or NCBC. The n.m.r. spectrum for a typical NCMC   shows

signals attributable to the N-carboxymethyl substituent, at
168.7 and 47.7 ppm, for COO$^-$ and N-CH$_2$, respectively. A down
field shift ($\sim$ 6 ppm) is observed for the carbon bearing the
N-carboxymethyl group, C-2', with respect to the corresponding carbon of unmodified residues, C-2 at 57.8 ppm. From
the ratio of the areas of signals C-2 and C-2' (as well as
the corresponding anomeric carbons C-1 and C-1'), N-acetylation is about 40 % and N-carboxymethylation more than 50 %.

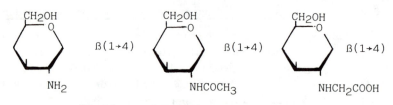

Residues of N-carboxymethyl chitosan

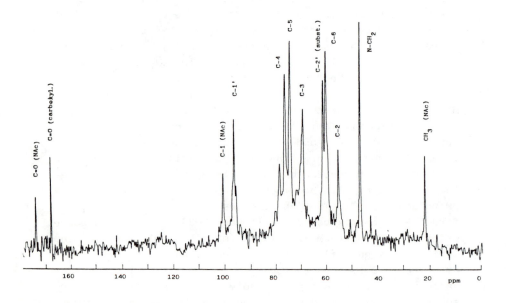

FIGURE 2. Residues present in N-carboxymethyl chitosan, and
the ^{13}C n.m.r. spectrum of a typical N-carboxymethyl chitosan
with degree of acetylation 40 %, degree of N-carboxymethylation
50 % and free amine 10 %. Muzzarelli, original results.

Some similarities exist between O-carboxymethyl chitosan and N-carboxymethyl chitosan in physical properties [50] (moisture content, fall in viscosity during the first few days after preparation, coagulation by solvents and neutral salts), but the primary alcohol groups of NCMC are available for further reactions such as cross-linking [51].

The attachment of reducing carbohydrates as side-chains to the 2-amino functions of chitosan transforms it into branched-chain water-soluble derivatives. Facile conversion can be achieved by reductive alkylation using sodium cyanoborohydride [45, 52]. Further, specific chemical modifications can be effected with the introduction, via D-galactose oxidase treatment, of aldehyde functions into the pendant D-galactose residue. Reductive alkylation of chitosan with lactose affords 1-deoxylactit-1-yl chitosan whose solution has unusual rheological properties.

7. MEDICAL AIDS

The wound-healing acceleration properties of chitin and chitosan alone and in cobination with other biopolymers are well assessed in the literature [2,12]. The degree of acceleration of the wound-healing process was determined in animal tests by measuring the bursting strength of the newly formed tissue of the wounds. The acceleration of healing was confirmed in human clinical tests.

Chitosan occludes arteries when injected in order to close arterio-venous fistulae and to infarct tumors; it prevents fibroplasia in vivo and in tissue cultures, thus permitting regeneration of normal tissue elements, and it also accelerate the growth of tissue in culture as well as encouraging it to grow in multi-level configuration, rather than in a monolayer [53].

A polyelectrolyte complex made of ammonium keratinate and chitosan acetate, with added collagen acetate can be prepared in the form of hydrogel membranes suitable as biodegradable wound dressings. Such membranes are uniformly and strongly adherent to wound tissues by virtue of their collagen content and consequent linkage to fibrin. They have a high degree of absorbancy for exudates and transport vapor at a rate sufficiently high so as to prevent fluid pooling

beneath the dressing and yet sufficiently low so as to maintain the desired moisture at the wound surface. These membranes are highly permeable to oxygen, allowing air to get
into the wound while stopping bacteria.   Their   microbial
barrier function can be further improved   by   incorporating
antimicrobial agents. Chitosan, keratin   and   collagen   are
non-antigenic and non-toxic. The polyelectrolyte complex is
storable after autoclaving and gas,   alcohol   and   radiation
sterilization. The moist membrane becomes biodegraded by the
skin cells, white cells and macrophages. In the latter stage
of wound healing, the remaining membrane hardens and falls
off without leaving any scarring [54].

The sulfation of N-(carboxymethyl) chitosan prepared
from Euphausia superba chitosan affords a glycosaminoglycan
carrying functional groups similar to those in heparin, and
its activity as a blood anticoagulant has been examined. The
results confirmed activity by the inhibition of thrombin by
binding to antithrombin and through inhibition of the factor
X_a-catalyzed conversion of prothrombin to thrombin [55, 56].
No appreciable hemolysis was produced and no adverse effects
were observed on lymphocytes or erythrocytes, as reported in
Table V.

TABLE V Morphological evaluations on human blood treated
with sulfated N-(carboxymethyl) chitosan (molecular weight
45,000; sulfur content 12.4 %). Muzzarelli, original results.

Erythrocytes	0 h	6 h	24 h
Morphology	++++	++++	++++
M. corpusc. vol.	++++	++++	++++
Sludge	absent	absent	absent
Hemolysis	absent	absent	absent
Platelets			
Morphology	++++	++++	++--
Sludge	absent	absent	absent
Activation	absent	absent	light
Polymorphonuclear neutrophils			
Morphology	++++	++++	++--
Activation	absent	absent	absent
Lymphocytes			
Morphology	++++	++++	++++

8. IMMOBILIZATION OF ENZYMES AND WHOLE CELLS

Immobilization techniques for living organisms and enzymes offer solutions to a variety of problems in biotechnology. Advantages can include: improvements in cell separation and product recovery, improved productivity of bioprocesses, continuous processes and improved cell stability. Table VI lists the enzymes which have been immobilized on chitin or modified chitins; most of them are of practical interest for industrial needs [57].

TABLE VI   Enzymes and cells immobilized on chitin or chitin derivatives, and proposed uses. Muzzarelli, Ref. 57 modified.

Amylase, diastase and glucoamylase: conversion of potato starch, raw corn starch and glycogen to D-glucose.
D-Glucose isomerase: isomerization of D-glucose to D-fructose
ß-D-Galactosidase: hydrolysis of lactose in milk and whey.
ß-D-Glucosidase: hydrolysis of cellobiose.
D-Glucose oxidase: preparation of D-gluconic acid.
AMP-deaminase: deamination of AMP to IMP.
Urease: conversion of urea to ammonia and carbon dioxide.
Papain: removal of haze from beer.
Pronase, subtilisin and trypsin: cosmetics and food proteins.
Pepsin: controlled digestion of proteins.
Chymotrypsin: plastein synthesis from soya and alfa-alfa.
Lysozyme: preparation of pharmaceuticals.
Invertase: sugar production.
Alkaline phosphatase and acid phosphatase.
Escherichia coli cells: synthesis of L-tryptophan.
Vibrio cholerae: epidemiological studies.
Nitrosomonas europaea cells: nitrification of waste water.

A current immobilization technique is based on the use of glutaraldehyde as a cross-linking agent to form aldimine bonds with both chitosan and protein. Direct immobilization can be performed with certain enzymes, for example gluco-amylase and α-amylase from the protease- and glucosidase-less mutant of Aspergillus awamori. Continuous production of high glucose concentrate from a concentrated α-amylase-treated gelatinized starch substrate was performed with the use of chitin-immobilized glucoamylase [58]. Other preferred supports are N-succinylated chitosan, N-succinylated glylcol chitosan, trimethylammonium glycol chitosan iodide, chitosan tripolyphosphate, chitosan-coated silica gel particles and chitosan alginate.

Novel organic-inorganic carriers compatible with fluidization have been proposed [59] for long term industrial operations, possessing high resistance to attrition and compression. Silica gel particles were coated with chitosan and used for the immobilization of trypsin, ß-galactosidase and invertase. For their preparation, 100 μm silica gel particle were treated with 1 % chitosan solution; after washing and drying, their chitosan coat was linked to the enzyme with glutaraldehyde or carbodiimide. Skim milk was passed through a column of the silica-chitosan-lactase at the flow-rate of 180 ml h^{-1} at 6.2°C in the counter-current mode: lactose was hydrolyzed nearly completely and the residual activity after 12 days was 85 %.

Entrapment methods are based on the inclusion of cells within a network: cells can not diffuse out of the network while nutrients can diffuse into the matrix. A variety of materials are widely used for entrapment processes, some examples being polyacrylamide gels, collagen, agar gels, alginates, cellulose triacetate and gelatin. Many of these immobilization processes involve severe conditions that may not be suitable for some biological systems. For example,the use of organic solvents, synthetic polymeric materials, high temperatures and extreme pH values may cause lysis or death of cells or damage their metabolic capabilities.In Table VII the cell loading capacities for several entrapment or immobilization methods are compared to those of chitosan matrix: high cell loadings are possible with the chitosan systems [60]. The growth rates of Bacillus subtilis and licheniformis were found to be dependent on the selected entrapment or encapsulation procedure. Cells in the chitosan alginate system were able to achieve growth rates comparable to those of free cells. The chitosan polyphosphate complex probably has a tighter network and growth was limited by reduced diffusion of materials to the cells.

TABLE VII   Cell loading capacity of various supports. Rha, Ref. 60, modified.

Support	Capacity, mg dry cell/ g support	Reference
Chitosan alginate	279	60
Chitosan polyphosphate	65 – 374	60
	376	61
Trimethylammonium glycol chitosan iodide with K poly(vinylalcohol) sulfate	83	62
Carrageenan	119	63
Synthetic polymers		
Dualite A-162	9	64
Dualite A-101	17	64
Dowex	21	64
Polyvinyl chloride	80	64
Inorganic supports		
Fritted glass	18.8	65
Cordierite ceramic	21.3	65
Zirconia ceramic	22.3	65

9. METAL ION CHELATION

The amino groups of chitosan are powerful ligands for binding to transition metal ions, and therefore chitosan better than chitin exhibits chelating activity for a number of metals, with the exclusion of alkali, alkali-earth   elements   and thallium. Early research on this peculiar characteristic of chitosan led to new methods of determination   of   transition metals in sea water, including Cu,   Mo   and   V,   and   to   the investigation of the stability of natural complexes existing in the marine environment. The naturally occurring complexes have to be oxidized before attempting   the   removal   of   metal ions with the aid of chelating polymers. Table VIII   shows

that after persulfate treatment of the natural complexes in
sea water, the cations can be quantitatively   removed   from
sea water and quantitatively isolated from exceedingly larger amounts of alkali and alkali-earth elements, by simply
percolating the sea water through a small chitosan column;
the transition metals are then eluted with dilute sulfuric
acid [15, 66, 67].

Aluminum can be collected on chitosan from lake waters:
it is in fact leached from the rocks by acid rain and it
represents a threat to aquaculture. Chitosan filters may be
an alternative to other water treatments [68].

A number of natural or man-made derivatives of chitosan
show enhanced chelating ability for   transition   metal   ions
than chitosan itself: among these, the chitosan-glucan complexes from Aspergillus niger, Mucor rouxii, Rhizopus arrhizus and others. Among the man-made derivatives of chitosan,
NCMC and epichlorohydrine cross-linked   glycine glucan,
obtained from fully deacetylated chitosan, are much more
powerful chelating agents than chitosan and chitosan-glucans
because their capacities are about 20 and 4 times higher,
respectively [69].

TABLE VIII   Recovery of metals from sea water, after persul
fate treatment and addition of known   amounts   of   metals.
Muzzarelli and Rocchetti, original results.

Metal	Sea water, untreated, $g \cdot l^{-1}$ as such,	with addition	Recovery of total amount, %
Cadmium	0.16 ± 0.02	1.15 ± 0.17	99
Lead	0.13 ± 0.03	0.83 ± 0.08	70
Nickel	0.94 ± 0.16	2.14 ± 0.54	60
Copper	2.64 ± 0.19	3.50 ± 0.42	43
	Sea water, treated, $g \cdot l^{-1}$		
Cadmium	0.16 ± 0.03	1.17 ± 0.33	101
Lead	0.20 ± 0.05	1.22 ± 0.07	102
Nickel	1.54 ± 0.19	3.44 ± 0.45	95
Copper	5.13 ± 0.75	7.05 ± 0.41	96

Glycine glucan, the NCMC obtained from fully deacetylated chitosan and glyoxylic acid, removes transition metal
ions from brines very effectively; the quantitative removal
of cobalt and copper at the $50 - 350$ μg l^{-1} concentration
is easily effected from sodium fluoride brines close to sa-
turation. The same metals at the concentration of $50 - 500$
μg l^{-1} are quantitatively removed from 13 % sodium chloride
brines. The data, relevant to solutions in the absence of
salts, are in Figures 3 and 4. The same Figures contain data
on aspartate glucan, which is also effective in chelating
transition metal ions: its capacities for cobalt and copper
are 18 and 80 mg g^{-1}, respectively, at equilibrium concentra
tions as low as 0.6 mg l^{-1}, [70].

Recently, cellulose derivatives containing amino acid
residues were tested for their chelating capacity; among
them, glycine cellulose, obtained from the isocyanato cellulose with glycine methyl ester in triethylamine and DMSO,
can be taken for comparison with glycine glucan obtained
from chitosan and glyoxylic acid. From the data in Table III
of Ref. 71, it can be calculated that glycine cellulose has
the capacity of about 1.6 mg Cu/g at equilibrium concentration 4.4 mg Cu/l, and about 0.7 mg Cu/g at equilibrium concentration 2.8 mg Cu/l. If we compare these data with those
in Figure 4, we can see that glycine glucan is about 31 and
60 times more powerful than glycine cellulose, under the con
ditions indicated; this also means that the slope of the iso
therm curve for glycine cellulose is greater than for glycine glucan, i.e. the latter works much better at low metal
ion concentrations. Thus, the straighforward reaction of
chitosan with glyoxylic acid provides the most effective
chelating polymer so far available.

The initial chelation of transition metal ions is probably followed by nucleation induced by the chelates and by
growth of aggregates of inorganic materials on the polymer
surface [72]. The chemical modifications of chitosan are there
fore intended to enhance the chelating ability by introducing suitable groups which, after chelation, promote the
deposition of metal oxides and hydroxides. The capacity of
about 1 g uranium per gram of the derivative obtained from
chitosan and dehydroascorbic acid (DTHC) should probably
be interpreted in this light because in the final product,

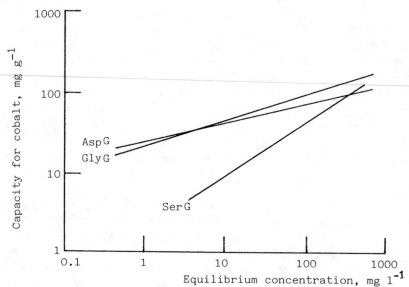

FIGURE 3.   Isotherms for the collection of cobalt on aspartate glucan (AspG), glycine glucan (GlyG) and serine glucan
(SerG). Muzzarelli et al., ref. 70.

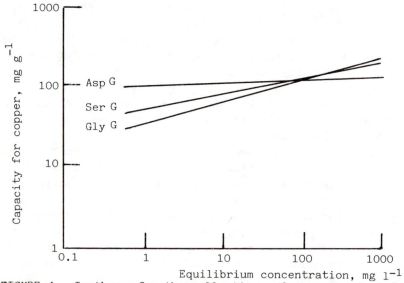

FIGURE 4.   Isotherms for the collection of copper. Symbols
as in Fig. 3. Muzzarelli et al., ref 70.

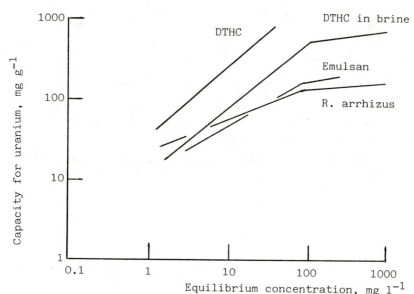

FIGURE 5. Isotherms for the collection of uranyl ion on the chitosan derivative obtained from dehydroascorbic acid (DTHC). The data are from references 73, 74 and 75.

the inorganic component is prevalent over the organic part [73]. Under comparable conditions, emulsan, the extracellular bioemulsifier of Acinetobacter calcoaceticus has capacity for uranium (about 200 mg g^{-1}, based on Figure 2 of Ref. 74) similar to the Rhizopus arrhizus treated mycelia capacity which was determined as 180 mg g^{-1}, [75]. As Figure 5 shows, it is evident that the novel chitosan derivative possesses much higher (about 1000 mg g^{-1}) capacity for uranium, and that it is more effective than the other products even at equilibrium concentration 2 mg l^{-1}. Incidentally, the emulsan backbone is composed of N-acetyl galactosamine, N-acetyl amino uronic acid and another unidentified amino sugar [76]. It is interesting to note that, among the other chemical derivatives of chitosan studied for uranium collection (over 30 modified chitosans) [77], the one obtained from dehydro-ascorbic acid is unique, because none was found to collect more uranium than chitosan itself.

REFERENCES

1. R. A. A. MUZZARELLI, Chitin (Pergamon, New York, 1977).

2. R. A. A. MUZZARELLI and E. R. PARISER (Eds.) Proceedings
   of the First International Conference on Chitin/Chitosan
   (M. I. T., Cambridge, 1978).

3. S. HIRANO and S. TOKURA (Eds.) Proceedings of the Second
   International Conference on Chitin/Chitosan (Japan Soc.
   Chitin, Tottori, 1982).

4. R. A. A. MUZZARELLI, in The Polysaccharides, edited by
   G.O. Aspinall (Academic Press, New York, 1984), Vol. 3.

5. H. R. HEPBURN (Ed.) The Insect Integument (Elsevier,
   Amsterdam, 1977).

6. R. C. W. BERKELEY, G. W. GOODAY and D. C. ELLWOOD (Eds.)
   Microbial Polysaccharides and Polysaccarases (Academic
   New York, 1979).

7. V. GINSBURG and P. ROBBINS (Eds.) Biology of Carbohydrates (Wiley, New York, 1981).

8. A. G. RICHARDS, in Biochemistry of Insects (Academic,
   New York, 1978), pp. 205-232.

9. G. W. GOODAY and A. P. J. TRINCI, in The Eukaryotic Microbial Cell, edited by G. W. Gooday, D. Lloyd and A.
   P. J. Trinci (Cambridge University, Cambridge, 1980),
   pp. 207-257.

10. J. H. BURNETT and A.P. J. TRINCI (Eds.) Fungal Walls and
    Hyphal Growth (Cambridge University, Cambridge, 1979).

11. J. F. KENNEDY and C. A. WHITE, Bioactive Carbohydrates,
    (Ellis Horwood, Chichester, 1983).

12. R. A. A. MUZZARELLI, Carbohydr. Polymers, 3, 53 (1983).

13. J. F. V. VINCENT, Structural Biomaterials (Macmillan,
    New York, 1982).

14. R. M. BROWN (Ed.) Cellulose and Other Natural Polymer
    Systems (Plenum, New York, 1982).

15. R. A. A. MUZZARELLI, Natural Chelating Polymers (Perga-
    mon, New York, 1973).

16. R. A. A. MUZZARELLI, F. TANFANI and G. SCARPINI, Bio-
    technol. Bioengin., 22, 885 (1980).

17. R. A. A. MUZZARELLI, F. TANFANI, G. SCARPINI and G. LA-
    TERZA, J. Biochem. Biophys. Methods, 2, 299 (1980).

18. R. A. A. MUZZARELLI and R. ROCCHETTI, to be published.

19. H. L. C. MEUZELAAR, J. HAVERKAMP and F. D. HILEMAN,
    Pyrolysis Mass Spectrometry, (Elsevier, Amsterdam, 1982).

20. D. GAGNAIRE, J. SAINTGERMAIN and M. VINCENDON, Makromol.
    Chem., 183, 593 (1982).

21. C. KIENZLE-STERZER, D. RODRIGUEZ-SANCHEZ and C. K. RHA,
    J. Appl. Sci, 27, 4467 (1982).

22. C. KIENZLE-STERZER, D. RODRIGUEZ-SANCHEZ, D. KARALEKAS
    and C. K. RHA, Macromol., 15, 631 (1982).

23. C. KIENZLE-STERZER, G. BAKIS, D. RODRIGUEZ-SANCHEZ and
    C. K. RHA, Polymer Bull., 11, 185 (1984).

24. M. BERGER, C. H. BYERS and J. J. MAGERHEIMER, U.S. Pat.
    4,386,151 (1983).

25. M. BERGER, U.S. Pat. 4,383,022 (1983).

26. R. A. A. MUZZARELLI, F. TANFANI, M. EMANUELLI   and   S.
    GENTILE, J. Appl. Biochem., 2, 380 (1981).

27. T. YANG and R. R. ZALL, J. Food Sci., 49, 91 (1984).   .

28. F. PITTALIS, F. BARTOLI and G. GIOVANNONI,   Eur. Pat.
    Appl. 0077098 (1983).

29. R. A. A. MUZZARELLI, A. ISOLATI and A. FERRERO, Ion Ex-
    change Membr., 1, 193 (1974).

30. P. GROSS, E. KONRAD and H. MAGER, Parf. Kosmetik, 64,
    367 (1983).

31. H. YAMADA and T. IMOTO, Carbohydr. Res., 92, 160 (1981).

32. G. K. MOORE and G. A. F. ROBERTS, Intl. J. Biol. Macro-
    mol., 3, 292 (1981).

33. N. NISHI, J. NOGUCHI, S. TOKURA and H. SHIOTA, Polymer
    J., 11, 27 (1979).

34. S. HIRANO and Y. YAGI, Carbohydr. Res., 83, 103 (1980).

35. G. K. MOORE and G. A. F. ROBERTS, Intl. J. Biol. Macro-
    mol,, 4, 246 (1982).

36. K. KAIFU, N. NISHI and T. KOMAI, J. Polym. Sci.,   19,
    2361 (1981).

37. S. FUJII, H. KUMAGAI and M. NODA, Carbohydr. Res., 83,
    389 (1980).

38. O. SOMORIN, N. NISHI, S. TOKURA and J. NOGUCHI, Polymer
    J., 11, 391 (1979).

39. S. HIRANO and Y. OHE, Carbohydr. Polymers, 4, 15 (1984).

40. K. KURITA, H. ICHIKAWA, S. ISHIZEKI, H. FUJISAKI and Y.
    IWAKURA, Macromol. Chem., 183, 1161 (1982).

41. R. YAMAGUCHI, Y. ARAI, T. ITOH and S. HIRANO, Carbohydr.
    Res., 88, 172 (1981).

42. S. HIRANO and T. MORIYASU, Carbohydr. Res., 92, 323 (1981)

43. R. YAMAGUCHI, Y. ARAI, T. KANEKO and T. ITOH, Biotechnol. Bioengin., 24, 1081 (1982).

44. L. A. BERKOVICH, M. P. TSYURUPA and V. A. DAVANKOV, J. Polymer Sci., 21, 1281 (1983).

45. L. D. HALL and M. YALPANI, JCS Chem. Comm., 1153 (1980).

46. R. A. A. MUZZARELLI, F. TANFANI and M. EMANUELLI, J. Membrane Sci., 16, 295 (1983).

47. R. A. A. MUZZARELLI, F. TANFANI, M. EMANUELLI and S. MARIOTTI, Carbohydr. Res., 107, 199 (1982).

48. R. A. A. MUZZARELLI, F. TANFANI, M. EMANUELLI and S. MARIOTTI, Carbohydr. Polymers, 2, 145 (1982).

49. R. A. A. MUZZARELLI and F. TANFANI, Pure Appl. Chem., 54 2141 (1982).

50. J. KOSHUGI, Eur. Pat. Appl. 0018131 (1980).

51. R. A. A. MUZZARELLI, F. TANFANI, M. EMANUELLI, D. P. PACE, E. CHIURAZZI and M. PIANI, Carbohydr. Research, 126, 225 (1984).

52. M. YALPANI, L. D. HALL, M. A. TUNG and D. E. BROOKS, Nature, 302, 812 (1983).

53. W. G. MALETTE, H. J. QUIGLEY, R. D. GAINES, N. D. JOHNSON and W. G. RAINER, Ann. Thorac. Surg., 36, 55 (1983)

54. A. WIDRA, Eur. Pat. Appl. 0089152 (1983).

55. R. A. A. MUZZARELLI, F. TANFANI and M. EMANUELLI, Carbohydr. Res.

56. R. A. A. MUZZARELLI, in Polymers in Medicine, edited by G. Chiellini and E. Giusti (Plenum, New York, 1984), pp. 359-374.

57. R. A. A. MUZZARELLI, Enzyme Microb. Technol., 2, 177 (1980).

58. P. Q. FLOR and S. HAYASHIDA, Biotechnol. Bioengin., 25, 1973 (1983).

59. J. L. LEUBA, Eur. Pat. Appl. 0079595 (182).

60. C. K. RHA, in Proc. Conf. Biotechnology of Marine Polysaccharides (M.I.T. Sea Grant, Cambridge, 1984).

61. K. D. VORLOP and J. KLEIN, Biotechn. Letters, 3, 9 (1981)

62. E. KOKUFUTA, W. MATSUMOTO and I. NAKAMURA, J. Appl. Pol. Sci., 27, 2503 (1982).

63. I. CHIBATA, T. TOSA, K. YAMAMOTO, I. TOKOTA and Y. NISHIDA, Enzyme Eng., 3, 335 (1978).

64. G. DURAND and J. M. NAVARRO, Process Biochem., 13, 14

(1978).

65. R. A. MESSING, Ann. Rep. Ferment. Processes,   4,   105
    (1980).

66. R. A. A. MUZZARELLI, G. RAITH and O. TUBERTINI, J. Chro-
    matogr., 47, 414 (1970).

67. R. A. A. MUZZARELLI and O. TUBERTINI, Talanta, 16, 1571,
    (1969).

68. R. A. A. MUZZARELLI, to be published

69. R. A. A. MUZZARELLI, F. TANFANI   and   M.   EMANUELLI,   J.
    Appl. Biochem., 3, 322 (1981).

70. R. A. A. MUZZARELLI, to be published

71. T. SATO, K. KARATSU, H. KITAMURA and Y. OHNO, Sen-I Gak-
    kaishi, 39, 519 (1983).

72. R. MARUCA, B. J. SUDER and J. P. WIGHTMAN, J. Appl. Pol.
    Sci., 27, 4827 (1982).

73. R. A. A. MUZZARELLI, F. TANFANI and M. EMANUELLI, to be
    published

74. Z. ZOSIM, D. GUTNICK and E. ROSENBERG, Biotechnol. Bio-
    engin., 25, 1725 (1983).

75. M. TSEZOS and B. VOLESKY, Biotechnol. Bioengin., 23, 583
    (1981).

76. Z. ZOSIM, D. GUTNICK and E. ROSENBERG, Biotechnol. Bio-
    engin., 24, 281 (1982).

77. S. HIRANO, Y. KONDO and Y. NAKAZAWA,   Carbohydr.   Res.,
    100, 431 (1982).

PREPARATION AND CHARACTERIZATION OF ENZYMATICALLY DERIVED OLIGOSACCHARIDES AND SEGMENTS FROM GLYCOSAMINOGLYCANS

MARY K. COWMAN
Department of Chemistry
Polytechnic Institute of New York
Brooklyn, New York 11201 USA

The controlled digestion of glycosaminoglycans by endoglycosidases yields oligosaccharide fragments which, to a first approximation, are composed of regularly repeating disaccharide units. Deviations from the idealized repeating structures exist in the position and degree of sulfation, and in the configurational isomers of hexuronate residues. These irregularities may be of importance in polymer behavior. Oligosaccharides of more uniform and well defined structure can provide useful model structures for the analysis of glycosaminoglycan conformation and interactions. Techniques for the isolation and spectroscopic characterization of oligosaccharides are described.

1. INTRODUCTION

Glycosaminoglycans are anionic polysaccharides found widely distributed in animal connective tissues. The development of detailed structure - function relationships for these species has been limited in part by the natural chemical heterogeneity and the high molecular weight of the intact polymers or their protein conjugates. The following report

describes the isolation and characterization of oligosaccharides which may be used to model and to alter glycosaminoglycan behavior in solution.

Figure 1 shows the idealized structures of several
glycosaminoglycans. Each is an unbranched chain with an
alternating sequence of residues[1]. One of the sugar residues is a hexuronic acid. It is β-D-glucuronic acid (GlcUA)
in chondroitin 4-sulfate, chondroitin 6-sulfate, and
hyaluronic acid. In dermatan sulfate, it is predominately
α-L-iduronic acid (IdUA). The second sugar residue is an
N-acetylated hexosamine. It is N-acetyl-β-D-glucosamine
(GlcNAc) in hyaluronic acid, but N-acetyl-β-D-galactosamine
(GalNAc) in the chondroitin and dermatan sulfates. Except
for hyaluronic acid, the glycosaminoglycans are sulfated,
and occur naturally in covalent attachment to protein. The
proteoglycan structures will not be discussed here. Rather
we will focus on the intrinsic properties of the glycosaminoglycan chains, and oligosaccharides derived from them
by enzymatic degradation.

Hyaluronic acid appears to possess no deviations from
the repeating structure shown in Figure 1. The other
glycosaminoglycans have occasional alterations in the
position and degree of sulfation. Dermatan sulfate also
has some GlcUA residues in place of IdUA. Such variations
in structure are of unknown function, but may be important
in self-association or other intermolecular interactions[2].

2. ISOLATION AND CHEMICAL ANALYSIS OF OLIGOSACCHARIDES

The isolation of structurally defined oligosaccharides for
conformational study is most simply accomplished with
hyaluronic acid (HA). Multiples of the disaccharide

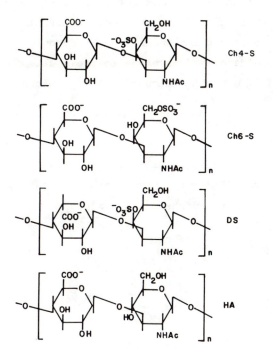

FIGURE 1   Structures of Chondroitin 4-Sulfate (Ch4-S),
    Chondroitin 6-Sulfate (Ch6-S), Dermatan Sulfate (DS),
    and Hyaluronic Acid (HA).

repeating structure may be obtained with either of two
hydrolytic enzymes. Bovine testicular hyaluronidase is an
endo β-hexosaminidase which cleaves HexNAc - GlcUA linkages
(HexNAc may be GlcNAc or GalNAc)[3,4]. The oligosaccharide
products of HA have the structure $GlcUA-(GlcNAc-GlcUA)_{n-1}-$
GlcNAc. A second, frame-shifted, set of oligosaccharides
may be obtained using leech head hyaluronidase. This endo
β-glucuronidase[5,6] cleaves HA to yield oligosaccharides with
the structure $GlcNAc-(GlcUA-GlcNAc)_{n-1}-GlcUA$.

The mixture of HA oligosaccharides generated by limited enzymatic digestion can be separated according to size
by gel filtration chromatography. Figure 2 shows the pro-
file obtained[7] using a Bio Gel P-30, minus 400 mesh, column
eluted with 0.5 M pyridinium acetate at pH 6.5. Species
containing up to approximately 15 disaccharide units are

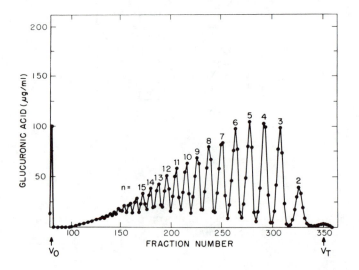

FIGURE 2   Gel filtration chromatography of HA oligo-
    saccharides on Bio Gel P-30, minus 400 mesh, eluted
    with 0.5 M pyridinium acetate, pH 6.5. From reference
    7, with permission.

separated into discrete peaks. The fine particle size of
the gel used in this case results in a slow flow rate. A
faster column of Bio Gel P-6, 200-400 mesh, eluted with the
same buffer, gives a nearly baseline separation of oligosaccharides containing up to 7-8 disaccharides in 2 days[8].
Since up to 100 mg of an oligosaccharide mixture may be
separated in a single chromatographic run, these methods
are suitable for the isolation of large quantities of
purified oligosaccharides. Following isolation, the oligosaccharide chain lengths may be determined by chemical
analysis. For bovine testicular hyaluronidase products,
the ratio of uronic acid residues[9] to reducing hexosamine[10]
is determined. For leech head hyaluronidase products, the
uronic acid loss after $NaBH_4$ reduction is monitored.

Enzymatic degradation followed by gel filtration
chromatography may also be used to isolate oligosaccharides
from the sulfated glycosaminoglycans. Chondroitin 4-sul-
fate (Ch4-S) and chondroitin 6-sulfate (Ch6-S) are susceptible to bovine testicular hyaluronidase[11,12]. Cleavage of
dermatan sulfate (DS) by this enzyme is limited to sites at
which GlcUA replaces IdUA[13-15]. In contrast to the case
for HA, however, the sulfated oligosaccharides isolated by
gel filtration chromatography are not homogeneous. Minor
variations remain in position and degree of sulfation, as
well as hexuronic acid type. Subsequent ion exchange
chromatography by either conventional[16-19] or high pressure
liquid chromatographic (HPLC) techniques[20] should be employed to obtain more uniform structures.

The arduous nature of large-scale oligosaccharide
isolation procedures has led to the development of methods
to rapidly analyze the content of enzymatic digests prior

to further treatment. HPLC methods have been developed for
this purpose[21,22]. An alternative technique is polyacrylamide gel electrophoresis. In our laboratory, we have used
a simple electrophoretic procedure[23] based on that used in
the separation of DNA fragments. For the separation of
relatively low molecular weight oligosaccharides (<20,000),
a 10 % polyacrylamide slab gel is employed (Figure 3).

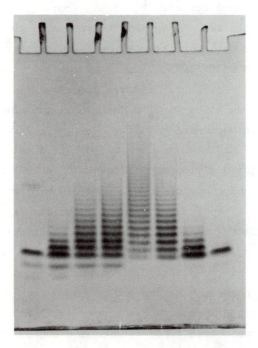

FIGURE 3   Polyacrylamide gel electrophoresis of sulfated
glycosaminoglycan oligosaccharide mixtures, produced
by digestion with bovine testicular hyaluronidase.
Lanes 1-4 from left: Ch6-S digested for 24 h, 2 h,
1 h, and 0.5 h, respectively. Lanes 5-8: Ch4-S digested for 0.5 h, 1 h, 2 h, and 24 h, respectively. Lane
1 also contains bromphenol blue tracking dye. From
reference 23, with permission.

The oligosaccharides present in the enzymatic digest of a
given glycosaminoglycan have approximately equal charge-to-
mass ratios, but are sieved to varying extents by the
polyacrylamide matrix. Thus the smallest oligosaccharides
migrate most rapidly through the gel. The separated oligosaccharides are visualized by soaking the gel in a solution
containing a cationic dye. Best results are obtained using
the multivalent copper phthalocyanine dye alcian blue to
precipitate and stain the oligosaccharide bands.

The identities of the stained bands have been determined by comparison with purified oligosaccharides.
Sulfated glycosaminoglycan oligosaccharides homogeneous
with respect to size were obtained by gel filtration on Bio
Gel P-6. Figure 4 shows the results of co-electrophoresis
of Ch6-S oligosaccharides with an unfractionated digest
sample. For the most part, each band in the electrophoretic pattern corresponds to an oligosaccharide species of
a unique chain length. Oligosaccharides containing 5 or
more disaccharide units appear as single bands, matching
the mobility of bands in the digest pattern. The smaller
oligosaccharides containing 3 or 4 disaccharide units show
two bands each. The extra bands are attributed to over-
sulfated species. In the smaller oligosaccharides, one
additional sulfate group is a sufficient perturbation of
the charge-to-mass ratio to result in a different electrophoretic mobility.

The screening of enzymatic digest samples may be
performed by simple visual inspection or densitometric
scanning of the stained gels. Figure 5 shows a comparison
of electrophoretic profiles and gel filtration patterns for
Ch4-S digested by bovine testicular hyaluronidase.

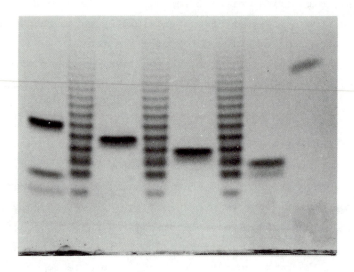

FIGURE 4   Identification of bands in the electrophoretic
       pattern of a Ch6-S digest, separated on 10% polyacryl-
       amide. Co-electrophoresis of oligosaccharides purified
       by gel filtration chromatography on Bio Gel P-6. Lanes
       2,4,6: Ch6-S digested 0.5 h with bovine testicular
       hyaluronidase. Lane 1: Tetradecasaccharide (n=7) and
       hexasaccharide (n=3). Lane 3: Dodecasaccharide (n=6).
       Lane 5: Decasaccharide (n=5). Lane 7: Octasaccharide
       (n=4). Lane 8: Bromphenol blue tracking dye. From
       reference 23, with permission.

There is an excellent correlation between the electrophoretic and chromatographic profiles, with one notable exception. The tetrasaccharide of Ch4-S (as well as Ch6-S and
DS) fails to stain under normal conditions.

    In summary, oligosaccharide fragments may be obtained
from polymeric glycosaminoglycans by digestion with
hydrolytic enzymes. Specific oligosaccharides may be
purified and analyzed by chromatographic and electrophoretic techniques. Samples of well defined covalent struc-

ture are suitable for use as models in conformational
studies.

3. OLIGOSACCHARIDE APPLICATIONS IN CONFORMATIONAL STUDIES

The solution conformation of high molecular weight HA has
been characterized as a somewhat stiff random coil[24],[25].

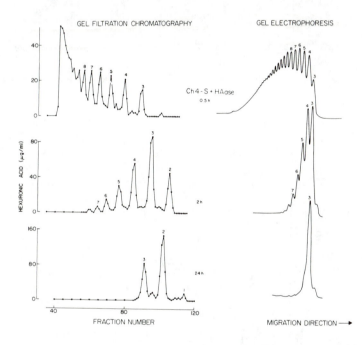

FIGURE 5   Comparison of Bio Gel P-6 gel filtration
    chromatographic profiles with densitometric scans of
    electrophoretic patterns for Ch4-S digested with
    bovine testicular hyaluronidase for varying periods of
    time.   The number of repeating disaccharide units in
    each species is indicated.   From reference 23, with
    permission.

One method which is applicable to the problem of providing
a greater degree of conformational detail is circular
dichroism (CD) spectroscopy. In neutral aqueous solution,
the CD spectrum of HA shows a negative band centered near
209 nm[7,8,26-28], as shown in Figure 6. This band has

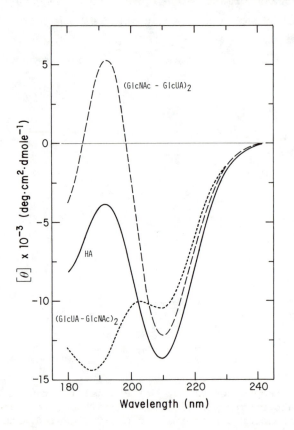

FIGURE 6   Vacuum ultraviolet circular dichroism spectra
      for neutral aqueous solutions of hyaluronic acid (HA)
      segment (15-20 disaccharides), and tetrasaccharides
      obtained after bovine testicular hyaluronidase
      [(GlcUA - GlcNAc)₂] or leech head hyaluronidase
      [(GlcNAc - GlcUA)₂] digestion. From reference 8, with
      permission.

contributions from the n-π* transitions of the amide and
carboxylate chromophores. In the 180-190 nm locations of
the π-π* transitions for these groups, no clear CD band is
noted. These spectral characteristics differ from those of
the monomeric GlcNAc and GlcUA constituents of HA. Moreover, tetrasaccharides derived from HA show large magnitude
CD bands near 190 nm, and differ from HA in the intensity
of the 209 nm band[8].

One plausible explanation for the difference in CD
properties of oligomeric and polymeric HA in aqueous solution invokes the existence of a helical polymer conformation
requiring a minimum number of repeating units for cooperative stabilization. Thus the small oligosaccharides would
be poor models for the polymer conformation. An alternative
explanation ascribes the differences more simply to the
covalent (and therefore spectroscopic) non-equivalence of
the end sugar residues of short oligosaccharide chains.
Changes in CD properties between unlinked monosaccharides,
singly-linked chain terminal residues, and fully-linked
internal residues would thus relate to local changes in
structure. To differentiate between these two explanations,
CD spectra were obtained for HA oligosaccharides containing
up to 8 disaccharide units[7,8]. The spectral properties of
oligosaccharides of either sequence were found to progress-
ively approach those of the polymer, as the oligosaccharide
chain length increased. A quantitative analysis of the data
indicated that the "end-effect" model adequately accounts
for the changes in CD band shape and intensity. The spectrum of each oligosaccharide can be fit as a linear combination of contributions from the end residues of the
chains, and the fully-linked internal residues of the chain.

The CD contribution of the internal residues of short
chains matches that of the polymer. Thus no evidence for a
cooperatively stabilized helical conformation was observed.

The CD contributions of the end residues in oligosaccharide chains were determined from the data obtained for the
larger species (Figure 7). For oligosaccharides with the
sequence GlcUA-(GlcNAc-GlcUA)$_{n-1}$-GlcNAc, the end residues
correspond in local structure to a β-1,3 linked GlcUA-GlcNAc
dimer. For the frame shifted oligosaccharides with the
sequence GlcNAc-(GlcUA-GlcNAc)$_{n-1}$-GlcUA, the end residues

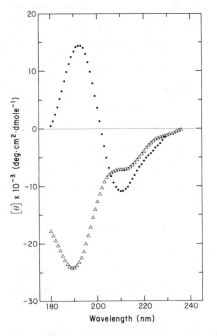

FIGURE 7   Calculated end residue contributions to the
circular dichroism spectra of HA oligosaccharides in
water. (O), end residues of the tetrasaccharide
GlcNAc-GlcUA-GlcNAc-GlcUA. (Δ), end residues of the
tetrasaccharide GlcUA-GlcNAc-GlcUA-GlcNAc. From
reference 8, with permission.

correspond to a β-1,4 linked GlcNAc-GlcUA dimer.   The
calculated spectral properties of these two structures are
distinctly different from each other, and from the polymer.
Large magnitude CD bands of opposite sign occur near 190 nm,
where the π-π* transition of the amide group is located.
Similar spectra have been previously observed[29,30] for model
compounds and glycoprotein oligosaccharides containing
acetamido sugars, but no uronic acid.   A β-glycoside of
GlcNAc can exhibit a large magnitude positive CD band near
190 nm.   A 3-linked GlcNAc can exhibit a large negative band
in this region.   Thus to a first approximation, the end
residue CD spectra are correlated with the linkage pattern
of the GlcNAc residue.

    A schematic illustration of the end residue structures
is helpful in interpreting these observations (Figure 8).
The planar amide group is optically active, in part as a
result of coupling with other chromophoric groups.   The
acetal chromophore is of particular importance.   In the two
end residue dimer structures, the acetal groups are located
in distinctly different positions with respect to the amide
group.   The opposite signs of the induced CD in the amide
π-π* bands may therefore reflect the relative orientations
of these chromophoric groups.   In the polymer, or in internal residues of smaller oligosaccharides, the GlcNAc residues are linked at both C1 and C3.   The observed low magnitude CD near 190 nm appears to result from offsetting
positive and negative contributions.

    It is worthwhile to pause at this point, and consider
whether any conformational detail concerning the structure
of HA in aqueous solution has been gained by CD analysis of
oligosaccharides.   The difference in CD properties of

polymeric and oligomeric HA has been attributed to the contributions of the singly-linked end residues in short chains
rather than to a major difference in chain conformation.
However, it is also possible to examine the data for end
residues to investigate the orientation of the amide group
in HA. The approach used[31] is one of analyzing the
perturbations in optical activity which accompany the
existence of glycosidic linkages. The difference in CD
properties between a GlcNAc - GlcUA dimer and the monomeric
sugar residues represents all changes in electrostatic
environment and interchromophoric coupling which occur with

D-GlcUA $\xrightarrow{\beta-1,3}$ D-GlcNAc

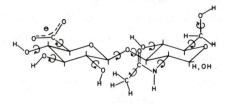

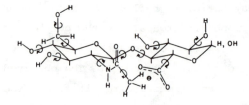

D-GlcNAc $\xrightarrow{\beta-1,4}$ D-GlcUA

FIGURE 8   Disaccharides which correspond in local
    structure to the end residues of HA oligosaccharides.

the formation of that β-1,4 linkage. Similarly, the difference between the GlcUA - GlcNAc dimer and the monomers
represents the changes which accompany the β-1,3 linkage
formation. The polymer CD spectrum may be directly compared to a sum of the monomer contributions and the two
linkage-dependent perturbations. It is found[32] to be a
good match in the spectral region of the amide and carboxylate group transitions. The additivity of perturbations
suggests that the orientation of the amide group with
respect to the sugar ring is not affected by glycosidic
linkage formation. Since the orientation of this group has
been established in the case of monomeric GlcNAc to be such
that the amide proton and the C2 ring proton are approximately _trans_, this same orientation is proposed to exist in
polymeric HA, in neutral aqueous solution.

The orientation of the amide group can also be investigated by nuclear magnetic resonance (NMR) spectroscopy.
The amide proton couples with the ring proton at C2. The
coupling constant is a monitor of the dihedral angle about
the C2-N bond. The amide proton has not been studied in
D_2O solution, due to its rapid rate of exchange, but has
been examined in H_2O solution[33]. For the monosaccharide
GlcNAc, both the α and β anomeric forms show well-resolved
amide proton resonances (Figure 9). The measured coupling
constants are approximately 9-10 Hz, indicating a nearly
trans arrangement of the ring and amide protons. A similar
analysis of the amide proton resonances for HA oligosaccharides and a larger HA segment also gives coupling constants
of 8-10 Hz at neutral and acidic pH, for all observed
resonances. These data establish the constancy of the
amide group orientation.

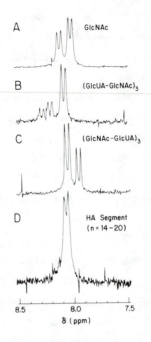

FIGURE 9   250 MHz ^{1}H-NMR spectra showing the amide proton
    resonances for HA segments and related models in water
    at pH 2.5. A, N-acetyl- D -glucosamine. The two
    resonances correspond to the α (upfield) and β (down-
    field) anomers. B, The hexasaccharide derived from HA
    by bovine testicular hyaluronidase digestion. The
    more intense (upfield) resonance corresponds to the
    interior GlcNAc residues. The downfield resonances
    are due to the α and β anomeric forms of the reducing
    end GlcNAc. C, The hexasaccharide derived from HA by
    leech head hyaluronidase digestion. The downfield
    resonance corresponds to the interior GlcNAc residues,
    and the upfield resonance is that of the nonreducing
    end GlcNAc. D, HA segment, 14-20 disaccharides in
    length. From reference 33, with permission.

The chemical shifts of the amide proton resonances
observed for HA oligosaccharides provide further information on the local environment of the amide group. Multiple

resonances are observed for small oligosaccharides. Analysis of the peak areas shows the resonances correspond to internal GlcNAc vs. chain terminal GlcNAc residues. The C3 linked reducing terminal GlcNAc residues show resonances for the α- and β-anomeric forms, slightly downfield of the resonances for monomeric GlcNAc. The Cl(β) linked non-reducing terminal GlcNAc residues show a single resonance which is slightly upfield of the monomer β-GlcNAc peak. Where both Cl and C3 linkages are present in interior GlcNAc residues, the resonance appears at an intermediate chemical shift. Thus the oligosaccharide data provide a fragmentation of the contributions to the chemical shift of the amide protons in the larger HA segment. In no case was a substantial downfield shift of an amide proton resonance observed. These data differ from the observations of Scott and coworkers[34-36], who have studied the amide proton NMR for HA oligosaccharides in dimethyl sulfoxide solution. Their data indicate the existence of a hydrogen bond from the amide NH to a carboxylate oxygen of the adjacent uronic acid residue across the β-1,4 linkage. In aqueous solution, the amide proton chemical shift and coupling constant indicate an absence of changes which would be characteristic of hydrogen bonding to the carboxylate group.

4. ADVANTAGES OF HIGHER MOLECULAR WEIGHT OLIGOSACCHARIDES

While short oligosaccharides of HA have proven extremely useful in the analysis of the CD and NMR properties of the polymer in salt-free aqueous solution, there are some aspects of the polymer behavior which they model poorly. Solutions of high molecular weight ($>10^6$) HA in moderate ionic strength solution show viscoelastic behavior,

indicative of a constantly forming and breaking network of
interacting chains[37-40]. Welsh et al.[39] have shown that HA
oligosaccharides containing 2-4 disaccharide units have no
effect on the polymer network, but a preparation of larger
average size (∿60 disaccharides) has the ability to reduce
the coupling between polymer molecules. These data suggest
that the network properties of HA are related to the formation of dimeric interaction zones. They further suggest
that only larger oligosaccharides are capable of interchain
association, and useable in efforts to modify the properties
of polymer solutions for commercial applications.

An additional property of HA oligosaccharides which
has been found to be strongly chain length dependent is
observed by polyacrylamide gel electrophoresis.[41]
Oligosaccharides of HA are separated according to size by
the sieving effect of the polyacrylamide gel, in the same
manner as observed for chondroitin and dermatan sulfate
fragments. The binding of cationic dyes and consequent
precipitation in the gel, however, depend dramatically on
the chain length for HA species. Oligosaccharides containing less than 7 repeating disaccharide units are not stained
by any of a wide variety of cationic dyes. Larger oligosaccharides show an increasing ability to bind dye. Constant maximal dye binding requires an oligosaccharide of
approximately 12 to 30 disaccharide units, depending on the
dye structure and solvent conditions. Two explanations
have been proposed for this phenomenon. The first considers
the change in dye binding to coincide with the transition
from simple electrolyte to polyelectrolyte behavior, with
increasing chain length. The second explanation proposes
the binding of cationic dyes to be associated with the

formation of HA interchain interactions, and the consequent
increase in effective linear charge density within the
interaction zones. In either case, oligosaccharides
greater than 7 repeating units in length may prove to be
superior models for the polymer conformation. The properties of ion binding and interchain interactions will be
best modelled by oligosaccharides of sufficient size to show
polymer-like behavior, while minimizing the possibilities
of intrachain association between distant segments of the
same chain, and the consequent complications in analysis
of physical data.

(This work was supported in part by Grant number EY 04804
from the National Institutes of Health.)

REFERENCES

1. R. W. JEANLOZ, in The Carbohydrates: Chemistry and
    Biochemistry, Vol. IIB, edited by W. Pigman and
    D. Horton (Academic Press, New York, 1970), pp. 589-
    625.
2. L.-A. FRANSSON, I. CARLSTEDT, L. COSTER, and
    A. MALMSTROM, J. Biol. Chem., 258, 14342-14345, (1983).
3. B. WEISSMANN, K. MEYER, P. SAMPSON, and A. LINKER,
    J. Biol. Chem., 208, 417-429, (1954).
4. B. WEISSMANN, J. Biol. Chem., 216, 783-794 (1955).
5. A. LINKER, P. HOFFMAN, and K. MEYER, Nature (London),
    180, 810-811, (1957).
6. A. LINKER, K. MEYER, and P. HOFFMAN, J. Biol. Chem.,
    235, 924-927 (1960).
7. M. K. COWMAN, E. A. BALAZS, C. W. BERGMANN, and
    K. MEYER, Biochemistry, 20, 1379-1385 (1981).
8. M. K. COWMAN, C. A. BUSH, and E. A. BALAZS, Biopolymers,
    22, 1319-1334 (1983).
9. E. A. BALAZS, K. O. BERNTSEN, J. KAROSSA, and
    D. A. SWANN, Anal. Biochem, 12, 547-558, (1965).
10. J. L. REISSIG, J. L. STROMINGER, and L. F. LELOIR,
    J. Biol. Chem., 217, 959-966. (1955).
11. K. MEYER and M. M. RAPPORT, Arch. Biochem., 27, 287-293,
    (1950).

12. M. B. MATHEWS, S. ROSEMAN, and A. DORFMAN, J. Biol. Chem., 188, 327-334, (1951).

13. L.-A. FRANSSON and L. RODEN, J. Biol. Chem., 242, 4161-4169, (1967).

14. L.-A. FRANSSON and L. RODEN, J. Biol. Chem., 242, 4170-4175, (1967).

15. L.-A. FRANSSON and A. MALMSTROM, Eur. J. Biochem, 18, 422-430, (1971).

16. L.-A. FRANSSON, J. Biol. Chem., 243, 1504-1510, (1968).

17. L.-A. FRANSSON, Biochim. Biophys. Acta, 156, 311-316, (1968).

18. A. MALMSTROM and L.-A. FRANSSON, Eur. J. Biochem, 18, 431-435, (1971).

19. M. HOOK, U. LINDAHL, and P.-H. IVERIUS, Biochem, J., 137, 33-43 (1974).

20. S. R. DELANEY, H. E. CONRAD, and J. H. GLASER, Anal. Biochem., 108, 25-34, (1980).

21. P. J. KNUDSEN, P. B. ERIKSEN, M. FENGER, and K. FLORENTZ, J. Chromatogr., 187, 373-379, (1980).

22. P. NEBINGER, M. KOEL, A. FRANZ, and E. WERRIES, J. Chromatogr., 265, 19-25, (1983).

23. M. K. COWMAN, M. F. SLAHETKA, D. M. HITTNER, J. KIM, M. FORINO, and G. GADELRAB, Biochem. J., 221, 707-716, (1984).

24. E. A. BALAZS, Fed. Proc., 17, 1086-1093, (1958).

25. T. C. LAURENT, in Chemistry and Molecular Biology of the Intercellular Matrix, Vol. 2, edited by E. A. Balazs, (Acad. Press, N. Y., 1970), pp. 703-732.

26. A. L. STONE, Biopolymers, 10, 739-751, (1971).

27. B. CHAKRABARTI and E. A. BALAZS, J. Mol. Biol., 78, 135-141, (1973).

28. L. A. BUFFINGTON, E. S. PYSH, B. CHAKRABARTI, and E. A. BALAZS, J. Am. Chem. Soc., 99, 1730-1734, (1977).

29. C. A. BUSH and S. RALAPATI, in Solution Properties of Polysaccharides, edited by D. A. Brant (American Chemical Society, Washington, D. C., 1981), pp. 293-302.

30. C. A. BUSH, R. E. FEENEY, D. T. OSUGA, S. RALAPATI, and Y. YEH, Int. J. Pept. Protein Res., 17, 125-129, (1981).

31. W. C. JOHNSON, JR., Carb. Res., 58, 9-20, (1977).

32. M. K. COWMAN, in Glycoconjugates, edited by M. A. Chester, D. Heinegard, A. Lundblad, and S. Svenson (Secretariat, 7th Int. Glycoconjugate Symposium, Lund, Sweden, 1983), pp. 61-62.

33. M. K. COWMAN, D. COZART, K. NAKANISHI, and
    E. A. BALAZS, Arch. Biochem, Biophys., 230, 203-212,
    (1984).
34. J. E. SCOTT, F. HEATLEY, D. MOORCROFT, and
    A. H. OCAVESEN, Biochem. J., 199, 829-832, (1981).
35. F. HEATLEY, J. E. SCOTT, R. W. JEANLOZ, and
    E. WALKER-NASIR, Carb. Res., 99, 1-11, (1982).
36. J. E. SCOTT, F. HEATLEY, and W. E. HULL, Biochem. J.,
    220, 197-205, (1984).
37. E. A. BALAZS, Fed. Proc., 25, 1817-1822, (1966).
38. D. A. GIBBS, E. W. MERRILL, K. A. SMITH, and
    E. A. BALAZS, Biopolymers, 6, 777-791, (1968).
39. E. J. WELSH, D. A. REES, E. R. MORRIS, and
    J. K. MADDEN, J. Mol. Biol., 138, 375-382, (1980).
40. E. R. MORRIS, D. A. REES, and E. J. WELSH, J. Mol.
    Biol., 138, 383-400, (1980).
41. R. E. TURNER and M. K. COWMAN, Arch. Biochem. Biophys.
    in press.

N E W   D R U G S   F R O M   H E P A R I N

B. C A S U

Istituto di Chimica e Biochimica "G. Ronzoni"
G. Colombo, 81
20133 Milano, Italy

Heparin has several biological activities, mostly as
sociated with interactions with plasma proteins and/or
components of the endothelium.   In spite of being
largely made up of regular sequences of a trisulfated
disaccharide, heparin chains are heterogeneous as re
gards sequences and size.   Some of the biological
activities of heparin are associated with different
structural features, and can be "concentrated" by phys
ical separation of chain families.   Specific activi
ties can be enhanced or suppressed by chemical or en
zymic modifications, including controlled depolymeriza
tion.

1. INTRODUCTION

Heparin, a sulfated polysaccharide extracted from animal

organs, has several biological activities, some of which

are at the basis of its use in therapy as an antilipemic

(fat clearing) and an anticoagulant (antithrombotic).[1]

The major clinical uses of heparin are based on the preven

tion of thrombosis in cardiovascular surgery and thromboem

bolism.[1,2]   Heparin is being evaluated also as a more ge

neral antiatherosclerotic drug (on account of its inhibito_
ry action on growth of arterial smooth-muscle cells)[3], as
an immunosuppressant (through inhibition of the complement
system)[4], and as an antitumor (through inhibition of forma_
tion of angiomas, regression of tumor masses and prevention
of metastases)[5].

   Most of the biological activities of heparin are associa_
ted with interactions with plasma proteins and/or components
of the endothelium.[5-8]   As illustrated in an idealized form
in Figure 1, the fat-clearing effect of heparin is caused
by release of the enzyme lipoproteinlipase from its cellular
receptors, including heparan sulfate (HS), a glycosaminogly_
can less sulfated than heparin.[8]

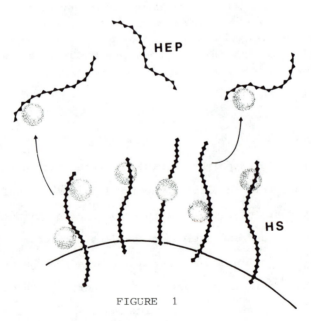

FIGURE   1

The anticoagulant properties of heparin are largely asso_
ciated with a more complex mechanism, involving inhibition
of enzymes at different levels of the "coagulation cascade",
either directly, or (more effectively) by enhanching the
effect of natural inhibitors such as Antithrombin-III.[10]
The enzymes of the coagulation cascade are produced from
inactive precursors as a reaction to damage of blood ves_
sels, and their concerted action rapidly leads to fibrin
clots.   As schematically illustrated in Figure 2, two
pathways converge to form Factor Xa, which in turn catalyzes
the formation of thrombin.[7,10]   In principle, thrombosis
can be prevented either by inhibiting the formation or the
activity of Factor Xa (thus preventing the formation of
thrombin), or by inhibiting thrombin once it is formed.
Because of the amplifying effect of the cascade, significan_
tly lower doses of heparin are required for the (antithrom_
bin-mediated) inhibition of Factor Xa than required for
thrombin.[2]

Though associated in part with the anticoagulant proper_
ties as measured in vitro from clotting of blood or thrombin
activity, the actual antithrombotic activity of heparin
(i.e., its ability to prevent formation of thrombi and/or
to lyse them once formed) is a result also of other proper_
ties, such as interaction with platelets, release of endo_
genous antithrombotic substances, and protection of the
surface of endothelial cells.[11]

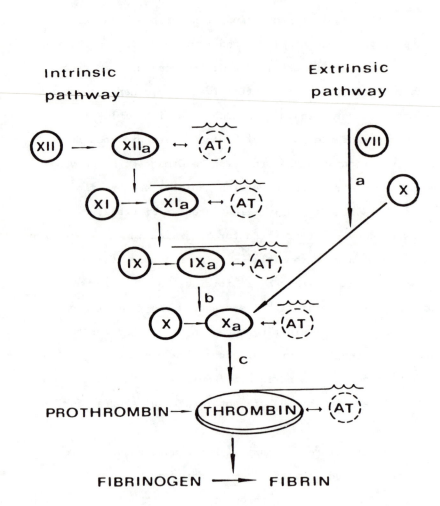

FIGURE 2. Simplified representation of the coagulation cascade.[8]   (Other co-factors, including calcium and phos‾ pholipids, are required in steps a, b and c. ; ──────⌒⌒⌒⌒ is a heparin chain, where ⌒⌒⌒⌒ is the binding site for antithrombin (AT).)

2. STRUCTURE OF HEPARIN. CHEMICAL AND BIOLOGICAL HETERO _
GENEITY

Like other glycosaminoglycans, heparin is a copolymer of
an uronic acid and a hexosamine, its structure being large_
ly accounted for by regular sequences of the trisulfated
disaccharide I_{2S}-$A_{NS,6S}$ (α-$\underline{L}$-iduronic acid 2-sulfate $\rightarrow$
2-deoxy-2-sulfamino-α-$\underline{D}$- glucose 6-sulfate). However, be_
cause of incomplete biosynthesis and also depending on the
history of the preparation, heparins from different animal
sources as well as from the same source and manufacturer,
contain different proportions of the foregoing sequences.
Typically, heparin preparations from the two most common
sources (i.e., beef lung and pig intestinal mucosa) contain
70-90% and 60-80%, respectively, of these sequences.
Irregular sequences contain nonsulfated uronic acids (either
$\underline{L}$-iduronic or $\underline{D}$-glucuronic) and $\underline{D}$-glucosamine residues la_
cking sulfate groups at position 6, or N-acetylated instead
of N-sulfated.   The heterogeneous regions of heparin also
contain specific sequences, such as the "linkage region"
and the antithrombin-binding sequence.[6-8]

   The linkage region, a tetrasaccharide segment containing
neutral sugars and terminating with serine, is the site of
attachment of heparin chains to the proteoglycan precursor
of heparin ("macromolecular heparin").   Not all the hepa_
rin chains necessarily terminate with the linkage region
sequence.   In fact, the polysaccharide chains of the ma_
cromolecular precursor are cleaved from the polypeptide

B. CASU

matrix and partially depolymerized by the concerted action
of peptidases and endoglycosidases.   The final product
of biosynthesis accordingly contains also chains termina_
ting with reducing carbohydrate residues.   Probably becau_
se of ramdom action of the glycosidases, and also depending
on irregular distribution of labile residues along the po_
lysaccharide chain of the heparin proteoglycan, the final
product of biosynthesis consists of chains of different si_
ze, usually ranging from 5,000 to 25,000.[12]   Such a size
heterogeneity can be further increased whenever drastic
conditions are used for extracting and purifying the pro_
duct.

The antithrombin-binding sequence is an oligosaccharide
segment containing $\underline{D}$-glucuronic acid (G)[13] as well as a
unique trisulfated $\underline{D}$-glucosamine residue ($A_{NS,6S}^{+}$).[14,15]
The "minimal" structure required for high-affinity binding
to antithrombin is a pentasaccharide,[16-18] usually linked
to a nonsulfated $\underline{L}$-iduronic acid residue,[13] probably acting
only as a biosynthetic marker for the specific sequence.
This sequence is contained in only about one third of the
chains constituting pig mucosal heparins, and apparently
even less for beef lung heparins.   The main structural
features of heparin are summarized in Figure 3.

The concept that the structural heterogeneity of heparin
could imply also biological heterogeneity (i.e., the possi
bility that different biological activities are associated
with different chains and/or chain segments) has stimulated

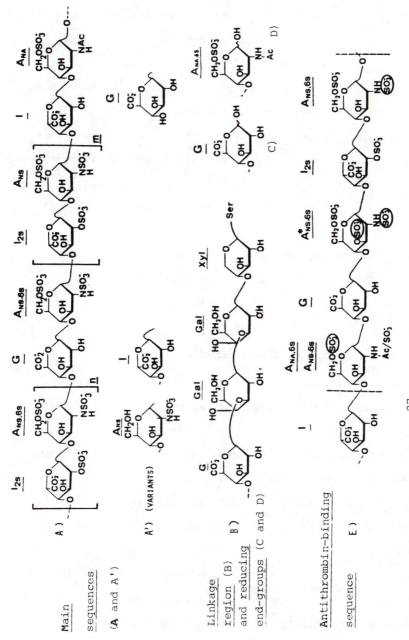

FIGURE 3.   Structure of heparin.[27]

extensive work on heparin fractionation and chemical and
enzymic modification, including fragmentation. A practi_
cal aim of these studies is to "concentrate" desirable bio_
logical activities, while "diluting" or removing undesira_
ble ones, such as unpredictable hemorrage. (Although hepa_
rin has a remarkable record as a nontoxic drug and its use
is regarded as essentially safe, either short-term use of
high doses or long-term treatment with sub-coagulant doses
may involve these hemorragic risks, as well as side effects
such as depletion of platelets (thrombocytopenia) or cal_
cium (osteoporosis.))[1]   Some of the biological properties
have been indeed shown to be associated with different
structural features, and enrichment of these structures
has been achieved in several instances.

3. FRACTIONATION OF HEPARIN CHAINS

Heparin chains can be fractionated according to specific
affinities for plasma proteins, as well according to size
(by gel filtration) and charge density (by precipitation
or partition methods, or ion-exchange). The most drama_
tic separation of activities was obtained by affinity chro_
matography (or precipitation) with antithrombin, with con_
centration of most of the anticoagulant activity in high-
affinity fractions,[19] consisting only of chains containing
the pentasaccharide sequence E) of Figure 3.[14,15] Although
some concentration was obtained also for the antilipemic
activity (by affinity chromatography on lipoproteinlipase),

fractions with lower affinity for the enzyme retained subs_
tantial antilipemic activity, all fractions being also anti_
coagulant.[20]

Fractionation of heparin by gel filtration has afforded
fractions of different average molecular weight, and permit_
ted to establish that the anticoagulant activity (as expres_
sed by generalized tests with whole blood, such as the U.S.P.
assay, or by measuring the activity of thrombin) significan_
tly decreases with decreasing size of the heparin chains,
[21,22] a trend confirmed also for chains with high affinity
for antithrombin.[23]   By contrast, inhibition of Factor Xa
is essentially independent of molecular weight.[21-23] This
behavior suggested that low-molecular weight heparin (LMW-
HEP) could be used for preventing venous thrombosis while
reducing the hemorragic risk usually associated with high
anticoagulant (thrombin-inhibiting) activities of unfractio_
nated heparin.[2]   Not unexpectedly, LMW fractions consis_
ting only of chains containing the active site for anti_
thrombin are much more effective than non-affinity fractio_
nated heparin in the antithrombin-mediated inhibition of
Factor Xa.[24]   Biological activities other than anticoagu_
lant (antilipemic, anticomplement) do not vary systemati_
cally with the molecular weight of heparin fractions.[25]

Separation of heparin species as a function of molecular
size can be obtained on a preparative scale by fractional
precipitation with alcohols of sodium or barium salts. (The_
se and other fractionation methods are reviewed in Ref.s

8, 26 and 27.)   However, fractionation by precipitation methods are affected also by charge density, the more sul_ fated species being less soluble in alcohols than less sul_ fated species of similar size.   This effect usually pre_ vents clear-cut separation of chain families, except in those (not unfrequent) cases where high molecular weights are associated also with high degrees of sulfation, in the presence of low molecular weight, low sulfate species. The same problem is encountered with methods based on polar interactions or ion-exchange.   Adsorption within the in_ terstices (cross-links) of resins causes molecular sieving effects which superimpose on charge interactions.   Advan_ tage can be taken of concurrent effects of charge and mo_ lecular sieving from excluding from basic resins the low-molecular weight/low sulfate heparin species.[28]

Fractionation of heparin based on charge density (inclu_ ding those based on affinity for basic proteins) has fre_ quently led to concentrate the anticoagulant activity in the most highly sulfated fractions, masking the role of the specific binding site for antithrombin.   This apparent contradiction can be explained by the current observation that the active site for antithrombin is most frequently inserted in heparin chains that are also longer and more sulfated than the average.   It should also be noted that development of full anticoagulant activity requires sequen_ ces of some disaccharides (essentially trisulfated) in addi_ tion to the active site for antithrombin,[29,30] this requi_

rement being probably associated with formation of ternary
complexes involving heparin, antithrombin and a coagulation
enzyme of the "thrombin series" (thrombin, Factor IXa and
Factor XIa), as schematically indicated in Figure 2.[31]
However, also chains with relatively low charge density
(and molecular weight) have affinity for antithrombin, and
indeed have high anti-Xa activity and contain the active
site for antithrombin.[32] Curves expressing the relation_
ship between charge density and anticoagulant activity are
"hinged" in the region of high activities,[33] most probably
because of predominance of the antithrombin-mediated mecha_
nism (requiring the specific active site for antithrombin)
over other mechanisms more generally associated with charge
density.

Charge density definitely plays a role in determining
biological interactions not mediated by antithrombin. The
"residual" anticoagulant activity of heparin chains that
not contain the active site for antithrombin (usually less
than 20% of the activity of unfractionated heparin) appears
to be due to direct inactivation of coagulation enzymes,
and/or to inactivation of these enzymes mediated by "Hepa_
rin Cofactor-2".[34,35] It is of interest that this cofac_
tor can be activated also by some of the chains of another
iduronic acid-containing glycosaminoglycan, i.e., dermatan
sulfate.[36]

4.   CHEMICALLY AND ENZYMICALLY MODIFIED HEPARINS

Specific activities of heparin can be modulated by chemical
or enzymic modifications. Partial removal of sulfate groups
impairs to some extent all the activities, with N-desulfa_
tion being especially effective in decreasing the anticoa_
gulant activity,[37] primarily by modifying critical residues
within the active site for antithrombin.   Complete (N- and
O-) desulfation[38] and carboxyl-reduction[39] remove the anti_
coagulant properties and all the biological properties bas_
ed on interactions involving charge density effects.   How_
ever, the anticomplement activity, which seems only a func_
tion of the degree of sulfation, is unmodified by complete
carboxyl-reduction.[39]   Esterification of carboxyl groups
modulates the interaction of heparin with different coagu_
lation enzymes.[37]

A major differentiation of biological activities is
achievable by selective cleavage (by periodate oxidation)
of the C(2)-C(3) bonds of nonsulfated uronic acid residues,
which dramatically reduces the anticoagulant activity[40]
while leaving the antilipemic activity substantially unaf_
fected or even enhanced.[41]   These findings are explainable
in terms of modifications within the active site for anti_
thrombin, with complete retention of the regular sequences,[41]
to which the lipoproteinlipase-releasing activity is as_
sociated.[42]

Biological activities can be further differentiated by

controlled enzymic or chemical depolymerization. Fragments
of the size of 10-30 monosaccharide residues (approximate
MW 3,000-9,000), essentially retaining the internal struc_
ture of the original chains (including the active site for
antithrombin) are most commonly obtained by partial cleava_
ge with heparinase[14,15] or nitrous acid.[14,15,29] As sche_
matically illustrated in Figure 4, the corresponding frag_
ments have different terminal residues. Base-catalyzed
depolymerization of heparin can afford fragments similar
to those produced by heparinase.[43] Internally non-modified
fragments can be obtained also by partial depolymerization
with acid, provided the acid-labile sulfamino groups are
restored by treatment with mild sulfating agents.[44] Under
strongly acidic conditions and in the presence of chloro_
sulfonic acid, sulfuric acid gives "supersulfated" LMW he_
parin fragments whose $\underline{\underline{D}}$-glucosamine residues are sulfated
also at O-3, as in the unique residue $A_{NS,6S}^{+}$ of the active
site for antithrombin.[45] (For a recent discussion on che_
mistry and biology of heparin fragments, see Ref. 46.)

Heparin fragments generally behave like LMW heparin frac_
tions of similar size in showing markedly reduced anticoa_
gulant activities as measured by whole-blood clotting as_
says or antithrombin activity, with substantial retention
of anti-Factor Xa activity.[15,29,30] However, high anti-X_a
to antithrombin activity ratios do not necessarily imply
high antithrombotic activities, at least in some experimen_
tal animals.[47,48] Since the capacity of plasma to generate

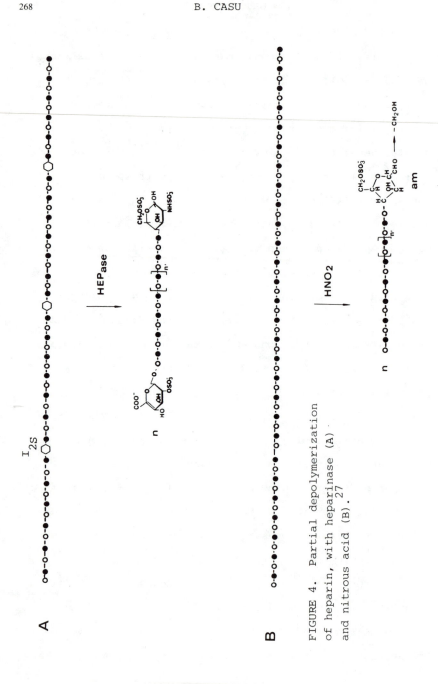

FIGURE 4. Partial depolymerization
of heparin, with heparinase (A)
and nitrous acid (B).[27]

thrombin seems to exceed the ability of heparin species to
enhance the inactivation of thrombin by plasma, only polys_
accharides that enhance this inactivation are likely to be
good antithrombotic agents in vivo.[49]   The antilipemic ac_
tivity of heparin fragments decreases with decreasing size
less steeply than the anticoagulant activity.[42,50]   The
increased density of sulfate groups as achieved in "super_
sulfated" heparin fragments, makes these fragments consis_
tently more antilipemic than unmodified heparin.[51]

Fragments smaller than decasaccharide are usually devoid
of biological activities, with the notable exception of
fractions[24] and fragments[14,15,29] containing the binding si_
te for antithrombin.   (These fragments, and similar synthe_
tic oligosaccharides,[18] have a high inhibitory activity on
Factor Xa, and potential antithrombotic activity.) Also,
hexasaccharide fragments obtained by cleavage of heparin
with heparinase were shown to retain antiangiogenic and an_
titumor properties of the unmodified preparations.[5]

5. FUTURE DEVELOPMENTS

Heparin is no longer regarded as a merely anticoagulant
drug, and its spectrum of applications in therapy is expan_
ding to prevention and treatment of various forms of throm_
bosis, atherosclerosis, inflammatory processes, and (poten_
tially) tumors.   Heparins for different uses will probably
be different from each other, as well as from today's unfrac_

tionated, unmodified preparations.   New forms of heparin
will most probably include:

1) Heparin with high affinity for antithrombin, i.e.,
consisting only of those polysaccharide chains containing
the active site for antithrombin.   2) Heparin with low af_
finity for antithrombin (prepared either by separation me_
thods, or by inactivation of the active site), for uses
other than anticoagulant and related therapies.   3) Low-
molecular weight heparin fractions and fragments, with
high anti-Xa and some antithrombin activity, for prophy_
laxis of venous thrombosis.   4) Heparins or heparin frag_
ments tailored for optimal inhibition of growth of smooth-
muscle cells, and of tumor cells, for use as antiatheros_
sclerotics and antitumors, respectively. (Structural re_
quirements for these activities, totally unrelated to anti_
thrombin, are still largely unknown.   The recent observa_
tion that a heparan sulfate fraction is much more effective
than heparin for inhibiting growth of smooth-muscle cells[52]
and that heparin fragments as small as a hexasaccharide (de_
void of other biological activities) is active in inhibiting
angiogenesis,[5] suggests that these effects are associated
with sequences somewhat different from those most recurrent
in "archetypal" heparin.   5) Preparations retaining their
activities when administered orally.   (Present commercial
heparins are not bioavailable orally, and a low molecular
weight is not necessarily the only requisite for absorption
through the intestinal wall.)   6) For all uses, heparins

involving <u>reduced hemorragic complications and side-effects</u>.

Chemical and biological standardization of the starting material is a prerequisite for the foregoing developments. Differences in biological properties associated with diffe_ rences in structure between the original heparin prepara_ tions could in fact be amplified in fractions, or modified species.   A major problem in the biological standardiza_ tion of heparin is the evaluation of actual hemorragic risks.   It is now recognized that these risks are not entirely reflected by the <u>in vitro</u> or <u>ex-vivo</u> anticoagulant activity, and have to be determined from <u>in vivo</u> bleeding tests.   Although some reports support the concept that LMW heparins cause less bleeding than unfractionated, un_ modified heparin,[53-55] others indicate unpredictable blee_ ding for some LMW preparations.[56,57]   Detection of a still unidentified heparin component in heparin preparations and fractions showing unusually high bleeding, suggests the existence of a specific "bleeding factor" that could indeed be responsible for occasional hemorragic phenomena when present in more than trace amounts.[57]   Identification and removal of this factor should greatly contribute to standar_ dization and safe use of heparin and LMW heparins.

Finally, it is now conceivable that some biological pro_ perties of heparin can be reproduced by total synthesis (as for the active site for antithrombin),[18,58] or by semi-syn_ tetic approaches simulating relevant sequences, with deve_ lopment of a new generation of "heparinoids".

REFERENCES

1. L. B. JAQUES, Heparins. Anionic Polyelectrolyte Drugs,
   Pharmacol. Rev., 31, 99-166 (1980).
2. D. P. THOMAS, Heparin, Clinics in Haematology, 10,
   443-458 (1981).
3. M. J. KARNOVSKY, Am. J. Path., 105, 200-206 (1981).
4. M.S. SY, E. SCHNEEBERGER, R. McCLUSLEY, M. I. GREENE,
   R. D. ROSENBERG, and B. BENACERRAF, Cell Immun., 82,
   23-32 (1983).
5. J. FOLKMAN, R. LANGER, R. J. LINHARDT, C. HAUDENSCHIELD,
   and S. TAYLOR, Science, 221, 719-725 (1983).
6. W. D. COMPER, Heparin and Related Polysaccharides
   (Gordon and Breach, New York, 1981).
7. I. BJÖRK and U. LINDAHL, Mechanism of the Anticoagu _
   lant Action of Heparin, Molecular Cellular Biochem,
   48, 161-182 (1982).
8. B. CASU, Structure and Biological Activity of Heparin,
   Advances Carbohydr. Chem. Biochem., in press.
9. T. OLIVECRONA, G. BENGSSTON, S.-E. MARKLUND, U. LIN_
   DAHL, and M. HÖÖK, Fed. Proc., 36, 60-65 (1977).
10. R. D. ROSENBERG, Fed. Proc., 36, 10-18 (1977).
11. L. B. JAQUES, Trends Pharmacol. Sci., 7, 289-291 (1982).
12. H. C. ROBINSON, A. A. HORNER, M. HÖÖK, S. ÖGREN, and
    U. LINDAHL, J. Biol. Chem., 253, 6687-6693 (1978).
13. R. D. ROSENBERG and L. LAM, Proc. Acad. Sci. USA,76,
    1218-1222 (1979).
14. J. HOPWOOD, M. HÖÖK, A. LINKER and U. LINDAHL, FEBS
    Letters, 69, 51-54 (1976); U. LINDAHL, G. BÄCKSTRÖM,
    M. HÖÖK, L. THUNBERG, L.-A. FRANSSON, and A. LINKER,
    Proc. Natl. Acad. Sci. USA, 76, 3198-3202 (1979).
15. B. CASU, P. ORESTE, G. TORRI, G. ZOPPETTI, J. CHOAY,
    J.-C. LORMEAU, M. PETITOU, P. SINAŸ, Biochem. J.,
    197, 599-609 (1981).
16. J. CHOAY, J.-C.LORMEAU, M. PETITOU, P. SINAŸ, and
    J. FAREED, Ann. N. Y. Acad. Sci., 370, 644-649 (1982).
17. L. THUNBERG, G. BÄCKSTRÖM, and U. LINDAHL, Carbohydr.
    Res., 100, 393-410 (1982).
18. J. CHOAY, M. PETITOU, J.-C. LORMEAU, P. SINAŸ,
    B. CASU, and G. GATTI, Biochem. Biophys. Res. Comm.,
    116, 492-499 (1983).

19. L. H. LAM, J. E. SILBERT, and R. D. ROSENBERG, Biochem. Biophys. Res. Comm., 69, 570-577 (1976); M. HÖÖK, I. BJÖRK, J. HOPWOOD, and U. LINDAHL, FEBS Letters,66, 90-93 (1976); L. O. ANDERSSON, T. W. BARROWCLIFFE, E. HOLMER, E. A. JOHNSON, and G. E. SIMS, Thromb. Res., 9, 575-583 (1976).

20. T. OLIVECRONA and G. BENGTSSON, in Intern. Conference Atherosclerosis, edited by. L. A. Carlson, R. Paoletti, C. R. Sirtori, and G. Weber (Raven Press, New York, 1978), pp. 153-157.

21. T. C. LAURENT, A. TENGBLAD, L. THUNBERG, M. HÖÖK, and U. LINDAHL, Biochem. J., 175, 691-701 (1978).

22. L. O. ANDERSSON, T. W. BARROWCLIFFE, E. HOLMER, E.A. JOHNSON, and E. SÖDERSTRÖM, Thromb. Res., 15, 531-541 (1979).

23. L. THUNBERG, U. LINDAHL, A. TENGBLAD, T. C. LAURENT, and C. M. JACKSON, Biochem. J., 181, 241-243 (1979).

24. J. CHOAY, J.-C. LORMEAU, M. PETITOU, P. SINAŸ, B. CASU, P. ORESTE, G. TORRI, and G. GATTI, Thromb. Res., 18 , 573-578 (1980).

25. P. BIANCHINI, B. OSIMA, B. PARMA, H. B. NADER, and C. P. DIETRICH, J. Pharmacol. Exp. Ther., 220, 406-410 (1982).

26. E. A. JOHNSON, Pharmacol. Res. Comm., 14, 289-320 (1982).

27. B. CASU, Nouv. Rev. Franç. Hémat., 26 (1984), in press.

28. B. CASU, G. TORRI and G. ZOPPETTI, US Patent 4,369,256 (1983); Carbohydr. Res., submitted.

29. G. M. OOSTA, W. T. GARDNER, D. L. BEELER, and R. D. ROSENBERG, Proc. Natl. Acad. Sci. USA, 78, 829-833 (1981).

30. E. HOLMER, U. LINDAHL, G. BÄCKSTRÖM, G. SÖDERSTRÖM, and L. O. ANDERSSON, Thromb. Res., 18, 861-869 (1980).

31. E. HOLMER, K. KURACHI, and G. SÖDERSTRÖM, Biochem. J., 193, 395-400 (1981).

32. B. CASU, J. CHOAY, J.-C. LORMEAU, M. PETITOU, and G. TORRI, unpublished results.

33. R. E. HURST, J. M. MENTER, S. S. WEST, J. M. SETTINE, and E. H. COYNE, Biochemistry, 18, 4283-4287 (1979).

34. F. A. OFOSU, G. MODI, A. L. CERKUS, J. HIRSH, and M. A. BLAJCHMAN, Thromb. Res., 28, 487-497 (1982).

35. R. E. HURST, M.-C. POON, and M. J. GRIFFITH, J. Clin.
    Inv., 72, 1042-1045 (1983).

36. D. M. TOLLEFSEN, C. A. PETSKA, and W. J. MONAFO,
    J. Biol. Chem., 258, 6713-6716; D. M. TOLLEFSEN, Ref.46.

37. I. DANISHEFSKY, Fed. Proc., 36, 33-35 (1977).

38. R. NAGASAWA, H. HARADA, S. HAYASHI, and T. MISAWA,
    Carbohydr. Res., 21, 420-426 (1972).

39. E. COFRANCESCO, F. REDAELLI, E. POGLIANI, N. AMICI,
    G. TORRI, and B. CASU, Thromb. Res., 14, 179-187 (1979).

40. L.-A. FRANSSON and W. LEWIS, FEBS Letters, 97, 119-
    123 (1979).

41. B. Casu, G. Diamantini, G. Fedeli, M. Mantovani,
    P. Oreste, R. Pescador, G. Prino, G. Torri, and G.
    Zoppetti, Arzneim.-Forsh., submitted.

42. B. CASU, E. A. JOHNSON, M. MANTOVANI, B. MULLOY,
    P. ORESTE, R. PESCADOR, G. PRINO, G. TORRI, and
    G. ZOPPETTI, Arzneim.-Forsch., 33, 135-142 (1983).

43. J. MARDIGUIAN and M. TRILLOU, Comm. IXth Int. Congress
    Thrombosis Haemostasis, Stockholm, 1983.

44. F. FUSSI and G. FEDELI, Ger. Offen. 2, 833-898 (1977).

45. A. NAGGI and G. TORRI , EP 116 801 (1984).

46. Proceedings Symposium Chemistry and Biology of Heparin
    Fragments, Nouv. Rev. Fr. Hematol., 26 (1984).

47. P. A. OCKELFORD, C. J. CARTER, L. MITCHELL, and J.
    HIRSH, Thromb. Res.,28, 401 (1982).

48. D. P. THOMAS, R. E. MERTON, T. W. BARROWCLIFFE, L.
    THUNBERG, and U. LINDAHL, Thromb. Haemostas., 47, 244 ,
    (1982).

49. F. A. OFUSU, M. A. BLAJCHMAN, M. R. BUCHANAN, G. J. MO_
    DI, L. M. SMITH, A. L. CERKUS, and J. HIRSH, unpublis_
    hed results.

50. G. BENGTSSON, T. OLIVECRONA, M. HÖÖK, J. RIESENFELD,
    and U. LINDAHL, Biochem. J., 189, 625-633 (1980).

51. B. CASU, J. LANSEN, A. NAGGI, and G. TORRI, unpublish_
    ed results.

52. R. D. ROSENBERG, in Ref. 46.

53. C. J. CARTER, J. G. KELTON, J. HIRSH, A. CERKUS,
    A.V. SANTOS, and M. GENT, Blood, 59, 1239-1245 (1982).

54. C. DOUTREMEPUICH, J. L. GESTREAU, M. O. MAURY,
    R. QUILICINI, M. R. BOISSEAU, F. TOULEMONDE, and
    E. VAIREL, Haemostasis, 13, 109-112 (1983).

55. V. V. KAKKAR, D. DJAZAERI, J. FOX, M. FLETCHER,
    M. F. SCULLY, and J. WESTWICK, Brit. Med. J., 284,
    375-379 (1983).

56. U. SCHMITZ-HUERNER, H. BÜNTE, G. FREISE, B. REES,
    C. RÜSCHEMEYER, R. SCHRERER, H. SCHULTE, and I. vande_
    LOO, Klin. Wochenschr., 62, 349-353 (1984).

57. M. ABBADINI, B. CASU, M. DONATI, A. NAGGI, J. PANGRAZ_
    ZI, I. REYERS, G. TORRI, and M. ZAMETTA, unpublished
    results.

58. J.-C. JAQUINET, M. PETITOU, P. DUCHAUSSOY, I. LEDERMAN,
    J. CHOAY, G. TORRI, and P. SINAŸ , Carbohydr. Res.,
    130, 221-241 (1984).

TOTAL SYNTHESIS OF BIOLOGICALLY ACTIVE
HEPARIN FRAGMENTS

M. PETITOU
Institut Choay
10, Rue Morel
92120 Montrouge, France

P. SINAŸ
Laboratoire de Biochimie Structurale
E.R.A. 739
U.E.R. de Sciences Fondamentales et Appliquées
45046 Orléans, France

In order to study the specificity of heparin-antithrombin III (AT-III) binding and to define its structural requirements, we have synthesized several oligosaccharides and have assessed their affinity for antithrombin III. It was found that the synthetic pentasaccharide (N-sulfate-6-O-sulfo-α-D-glucosamine)1→4(β-D-glucuronic acid) 1 →4(N-sulfate-3,6-di-O-sulfo-α -D-glucosamine) 1 → 4(2-O-sulfo- α -L-iduronic acid) 1 →4(N-sulfate-6-O-sulfo-α -D-glucosamine) strongly binds to AT-III and enhances the AT-III inhibitory activity towards factor Xa. The tetrasaccharide(β -D-glucuronic acid)1 → 4(N-sulfate-3,6-di-O-sulfo- α-D-glucosamine)1 → 4(2-O-sulfo- α-L-iduronic acid)1→4(N-sulfate-6-O-sulfo-D-glucosamine) neither binds to AT-III nor induces anti-factor Xa activity enhancement of this inhibitor. The tetrasaccharide (N-sulfate-6-O-sulfo- α -D-glucosamine)1 → 4(β -D-glucuronic acid)1→ 4(N-sulfate-3,6-di-O-sulfo-α -D-glucosamine)1 → 4(2-O-sulfo-L-iduronic acid) was able to elicit anti-factor Xa activity, although to a lesser extend than the pentasac-

charide. These resuts confirm that the synthetic
pentasaccharide with the above structure corresponds to the actual minimal sequence required
in heparin for binding to AT-III. The critical
role played by the unique 3-O-sulfate group
in the biological activity has also been demonstrated.

It is now fully recognized that the carbohydrate parts
of glycoconjugates-glycoproteins or glycolipids-play
a critical role in biology as they offer a unique archi-
tecture which is ideal for precise chemical recognition.
A good example is given by human blood group substances
of the ABO group, where the blood group specificity is
given by a trisaccharide[1]. A slight change in the structure of the molecule- the replacement of a secondary
hydroxy group by an acetamido group-completely modifies
the serological specificity. Organic synthesis- which
is often the only way to get an oligosaccharide in reasonable amount and with excellent purity- should therefore
be a technic of choice for studies of structure-activity
relationship in general. The purpose of this lecture
is to illustrate the importance of the chemical synthesis
of oligosaccharides for the investigation of the molecular
basis of the anticoagulant activity of heparin.

Heparin inhibits a number of procoagulant proteases
mainly by binding to antithrombin III and enhancing the
effects of this inhibitor[2]. Some ten years ago, the heparin-antithrombin III interaction was not known to be
specific and the role of charge density of the carbohydrate polymer was assumed to be important in relation
to anticoagulant activity[3]. However, it has been shown
that only about one third of the polysaccharide chains

in crude commercial heparin preparations have high affinity for antithrombin III, suggesting that the interaction
may be specific and may require very peculiar domains
of the polysaccharide chain. It is important to note
at this stage, that, with respect to the inhibition of
blood coagulation factors, they may be classified into
two types:factor Xa type and factor IIa[4] type. Whereas
high molecular weight heparin species are able to interact
with both types, a decrease in molecular weight is correlated with a decrease in activity towards the factor
IIa type[5-7]. An intense interest was thus devoted in
the past few years to the determination of the minimal
molecule weight heparin fractions still retaining high
anti-factor Xa activity in plasma. The structure of high-
affinity oligosaccharides, prepared from heparin by extraction, partial deaminative cleavage, or partial depo-
lymerisation with bacterial heparinase, have been studied[7,8-11]. An important result[12] was obtained in 1981
by J. Choay et al. when an active hexasaccharide was
prepared by exhaustive heparinase cleavage of high-affinity heparin fractions. Since it contained a modified
(4,5-unsaturated)uronic acid at its non-reducing end,
we hypothesized for the first time that, in heparin,
the minimal sequence that binds to antithrombin III and
elicits high anti-Xa activity was in fact the following
pentasaccharide :

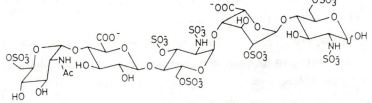

An experimental support[13] for this hypothesis was obtained by Lindahl's group after enzymatic removal of non-essential residues in species with antithrombin III high affinity. However, at this point, no pentasaccharide *per se* has been obtained in sufficient quantities for investigation.

In order to test this hypothesis and to investigate the precise structural requirements for specific inter-action between heparin and antithrombin III, Institut Choay and the Laboratoire de Biochimie Structurale (Univer-sité d'Orléans) started a synthetic program toward the pentasaccharide, various fragments, and structural ana-logs. For convenience, we chose to ultimately prepare the following tri-N-sulfated structural variant, as it occurs in bref-lung heparin :

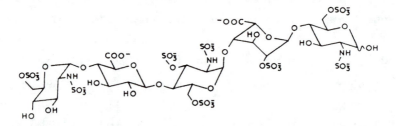

This program represents the first approach to syn-thesize specifically substituted glycosaminoglycan se-quences . Among other problems, it has required synthesi-zing α-linked L-iduronate residues and introducing sulfate groups into the appropriate positions in the L-iduronic and D-glucosamine residues.

The general strategy of this synthesis is based on benzyl ethers as permanent blocking groups. Although, in the past, problems have been encountered during the catalytic hydrogenolysis of benzyl ethers in the presence of sulfate groups, a preliminary study of monosaccharide models demonstrated that such a reaction was indeed possible.

Condensation of the orthoester **1** with the secondary alcohol **2** provided the protected disaccharide **3** (40%) which was converted into the alcohol **4** (86%). This alcohol reacts smoothly with the bromide **5** to give the trisaccharide **6** (88%). A series of reactions (i, NaOH, MeOH-H$_2$O; ii, H$^+$; iii, CH$_2$N$_2$; iv, Me$_3$N-SO$_3$, DMF; v, H$_2$, Pd/C; vi, Me$_3$N-SO$_3$, H$_2$O; vii, OH$^-$) provided **7** ,a fragment of the irregular segment of heparin[14].

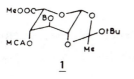

1

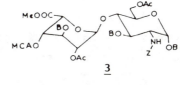

3

2

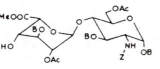

4

5

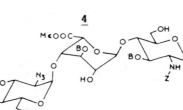

6

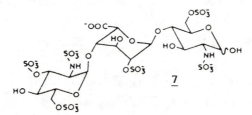

7

Antithrombin III-binding experiments by gel filtration and anti-Xa activity measured either by the clotting
assay of Yin et al. or by amidolytic assay of Teien
and Lie showed that the synthetic trisaccharide **7** neither
binds to antithrombin III nor induces anti-Xa activity .

These results indicate that, although being a part
of the postulated fragment of the heparin to antithrombin
III -binding site, the trisaccharide **7** does not contain
all the minimal structural requirements to bind antithrombin III and to induce anti-Xa activity[14].

The alcohol **4** was then successfully condensed with
the disaccharide bromide **8** to provide, in 80% yield,
the protected tetrasaccharide **9**, which was converted
as previously described into the tetrasaccharide **10**,
which again presented no affinity for antithrombin III

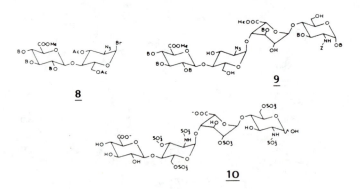

The alcohol **4** was now condensed with the disaccharide
bromide **11** to give, after de-monochloroacetylation ,
the alcohol **12** which was successfully condensed with
the known bromide **13** to provide the protected pentasaccharide **14**.

The series of deprotections and sulfations previously
reported gave the target pentasaccharide DEFGH[15] which

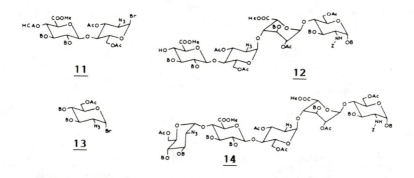

presented a high-affinity for antithrombin III. The complex between synthetic DEFGH and antithrombin III has an association constant of 7.10^6 M^{-1}, which is the same order of magnitude as that of high-affinity heparin. Furthermore, in the presence of antithrombin III, DEFGH displays a high-inhibitory activity with respect to factor Xa but, as expected, is unable to activate anti-thrombin III in the thrombin inhibition process[16].

It must be underlined that the previous results concerning pentasaccharide sequence and activity where deduced from studies performed on larger fragments in which the additional residues were assumed not to affect the activity. Even if this assumption looked valid, the chemically synthesized pentasaccharide DEFGH provides a remarkable answer to the question of the length of the oligosaccharide involved in the binding site.

This work demonstrates the importance of residue D for the binding to antithrombin III. As the 2-sulfamido group is non-essential[13], it is thus probable that the 6-sulfate plays an essential role.

In order to assess the role of the terminal gluco-samine unit H, we synthesized the following tetrasaccharide DEFG:

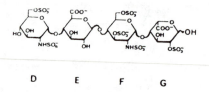

D   E   F   G

The orthoester **1** was condensed with benzyl alcohol to give, after selective O-dechloroacetylation, the alcohol **15** which was then transformed into DEFG following essentially previously disclosed methodology.

The synthetic tetrasaccharide DEFG was able to elicit anti-factor Xa activity[17], although to a lesser extend than DEFGH : the antifactor Xa activity (U/mg) was 600, compared to 4000 for DEFGH (Units refer to a house standard of low molecular weight calibrated with the 3rd international heparin standard). We thus can conclude from this synthetic work that the pentasaccharide DEFGH corresponds to the actual minimal sequence that is required in heparin commercial preparations to strongly bind antithrombin III and to induce high and specific anti-Xa activity.

An important feature of the use of organic synthesis in the field is that it allows the synthesis of analogs of defined structures, most of them being out of reach by biochemical or chemical modifications of the natural product. A typical illustration is provided by the chemical synthesis of the following pentasaccharide, which was performed by a similar route:

Compared to DEFGH, it lacks one ester sulfate group on the central glucosamine moiety. In ^{13}C-NMR analysis, this sulfate group induces a shift of the adjacent carbon--2 signal. This "extra signal" was first detected[7] in high antithrombin III affinity oligosaccharides and it was later shown to be associated with the heparin antithrombin binding sequence and, more precisely, assigned to the 3-0-sulfated glucosamine unit[11]. This modified pentasaccharide failed to elicit anti-factor Xa activity in human plasma and it did not bind to antithrombin III in qualitative experiments. This finding points to a critical role for the 3-0-sulfate group in pentasaccharide DEFGH. It is clear that selective oligosaccharide synthesis of various analogs of the active pentasaccharide DEFGH would result in a precise assessment of the sulfate groups which are essential for the biological activity of heparin.

Synthetic heparin fragments are not only useful for structure/activity relationship studies but as standards for biological and chemical determinations. Thus a compound with a well define structure might become a standard for anti-factor Xa assay.

Structural studies on heparin are nowadays carried out by sophisticated chemical and physicochemical methods that need standardization. Synthetic oligosaccharide samples are much useful in this respect. Thus, the chemical synthesis of monosaccharides and their analysis by ^{13}C-NMR spectroscopy has been helpful in the interpretation of data related to the antithrombin III binding sequence. In the same way an analysis of the ^{13}C- and ^{1}H-NMR spectra presented here (Figs 1 and 2) will allow

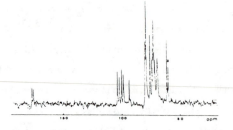

FIGURE 1 ^{13}C-NMR spectrum of pentasaccharide DEFGH.
The five signals at 103.75,102.21,100.31,98.92, and
93.91 ppm are assigned to anomeric carbons of units
E,G,D,F,and H respectively.Signal at 59.40 ppm (*) is
assigned to C-2 of 3-O-sulfated glucosamine unit F.

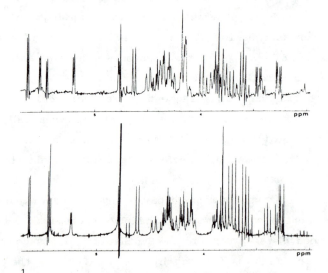

FIGURE 2 ^{1}H-NMR spectra of pentasaccharides DEFGH (upper
part) and DEFGH lacking the 3-O-sulfo group (lower part).
Careful analysis showed that the spectra are in agreement
with the expected structures.

the assignment of the different signals to the known structural features of the molecules. The same is true with M.S. spectra : figure 3 shows the mass spectrum (FAB) of the tetrasaccharide DEFG previously synthesized.

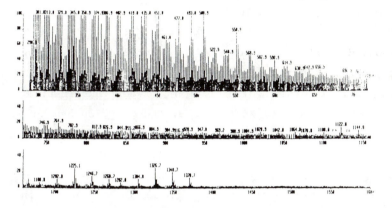

FIGURE 3 Mass spectrometry (FAB) spectrum of the tetra-saccharide EFGH (MW 1348 daltons).

We have emphasized here some of the applications of chemical synthesis to heparin. One could also add the use of such derivatives as enzyme substrates that would open the way to a better understanding of the structural requirements of such enzymes.

It is clear that selective organic synthesis of oligosaccharides is going to play a more and more important role in the development of the understanding of the molecular basis of biological activity of natural polysaccharides and glycoconjugates. Human blood group substances has recently[1] been an example; heparin is a timely one. We consider that the oligosaccharide fragments of plant polysaccharides are also potential targets[18]. Would it be possible to reach the stage where the design of a future drug or active substance of carbohydrate nature will be followed by bulk industrial production by synthesis? The power of combined chemical and biochemical technologies may be a possible answer to this challenging practical problem.

Acknowledgements. We are very grateful to Mr. Jean Choay as instigator of this work.The synthetic results arise from a close collaboration between the Institut Choay (Paris, France) (J.Choay, B. Chung, P. Duchaussoy, I. Lederman, J.-C. Lormeau, M. Petitou) and the Laboratoire de Biochimie Structurale (E.R.A. 739,Université d'Orléans, France) (J.-C. Jacquinet, P. Sinaÿ). Development of this work was also made in collaboration with Istituto Ronzoni (Milano, Italy) (B.Casu, G. Torri). We also thank G. Gatti (Brucker-Instruments, Milano, Italy) for is expert contribution to the NMR studies and V.N. Reinhold (Harvard University, Boston, U.S.A.) for the MS analysis.

REFERENCES

1. R.U. LEMIEUX, Frontiers of chemistry (Pergamon Press, Oxford, 1982), p.3.
2. R.D. ROSENBERG and P.S. DAMUS, J.Biol.Chem.,248, 6490-6505, (1973).
3. J.A. CIFONELLI, Carbohydr.Res., 37, 145-154, (1974).
4. E. HOLMER, K. KURACHI, and G. SÖDERSTRÖM, Biochem.J., 193, 395-400, (1981).
5. T.C. LAURENT, A. TENGBLAD, L. THUNBERG, M. HÖÖK, and U.LINDAHL, Biochem. J., 175, 691-701, (1978).
6. L.O. ANDERSSON, T.W. BARROWCLIFFE, E. HOLMER, E.A. JOHSON,and G. SODERSTROM, Thromb.Res.,15, 531-541, (1979).
7. J. CHOAY, J.C. LORMEAU, M. PETITOU, P. SINAŸ, B.CASU,

P. ORESTE, G. TORRI, and G. GATTI, Thromb.Res., 18,573-578,(1980).

8. R.D. ROSENBERG, G. ARMAND, and L. LAM, Proc.Natl.Acad. Sci.U.S.A., 75, 3065-3069, (1978).

9. U. LINDAHL, G. BÄCKSTRÖM, M. HÖÖK, L. THUNBERG, L.-Å.FRANSSON, and A. LINKER, Proc.Natl.Acad.Sci. U.S.A.,76, 3198-3202, (1979).

10. N. OTOTANI and Z. YOSIZAWA, J.Biochem. (Tokyo), 90, 1553-1556, (1981).

11. B. CASU, P. ORESTE, G. TORRI, G.ZOPPETTI, J. CHOAY, J.C. LORMEAU, M. PETITOU, and P. SINAŸ, Biochem.J., 197, 599-609, (1981).

12. J. CHOAY, J.C. LORMEAU, M. PETITOU, P. SINAŸ, and J. FAREED, Ann.N.Y.Acad.Sci., 370, 644-649, (1981).

13. L. THUNBERG, G. BÄCKSTRÖM, and U. LINDAHL, Carbohydr. Res., 100, 393-410, (1982).

14. J.C. JACQUINET,M.PETITOU,P.DUCHAUSSOY,I.LEDERMAN, J. CHOAY, G. TORRI, and P. SINAŸ, Carbohydr.Res., 130, 221-241, (1984).

15. P. SINAŸ, J.C. JACQUINET, M. PETITOU, P. DUCHAUSSOY, I. LEDERMAN, J. CHOAY, and G. TORRI, Carbohydr.Res., 132, C5-C9, (1984).

16. J. CHOAY,M.PETITOU, J.C.LORMEAU, P.SINAY,B.CASU, and G.GATTI, Biochem.Biophys.Res.,116,492-499,(1983).

17. M. PETITOU, Nouv.Rev.Fr.Hematol., 26, 221-226, (1984).

18. P. ALBERSHEIM, this workshop.

SOME COMMENTS ON COMMERCIAL PROCESSING

M. W. RUTENBERG
Natural Polymer Research Dept.
National Starch & Chemical Corp.
10 Finderne Avenue
Bridgewater, New Jersey 08807 U.S.A.

In these two days, we have heard fascinating reports of
some complex and elegant chemistry. This is quite far
removed from the relatively simple concepts of commercial
starch chemistry where the method of preparation and isolation of the product must result in yields and costs that
can compete in the marketplace and provide the quality and
combination of properties needed for the specific industrial or food application. To successfully scale-up a
laboratory preparation requires the cooperation and cre-
ativity of the research chemist and the process development
engineer. The chemistry that has been discussed at this
conference generally would be quite difficult to apply on
the commercial scale used in the manufacture of modified
starches where reactions are carried out in tanks that
hold in the neighborhood of 100,000 pounds of starch in
an aqueous suspension of about 30% solids. Remember also
that the purification of the starch product and disposal
of the effluent waste must be considered so as to meet
governmental regulations. Despite the comment made today
that "starch is a monotonous polysaccharide" (and I most

emphatically disagree), I do want to give some idea of the chemistry that is applied commercially.[1]

Cationic starches are manufactured and sold in the United States in the millions of pounds, primarily as wet-end additives in the manufacture of paper to provide strength, pigment and fines retention and good drainage properties.[1] One method of preparation involves treatment of an aqueous suspension of granular starch with N-(2-chloroethyl)-N,N-(diethyl)amine under alkaline conditions (pH 10-11.5).[2] As shown in the accompanying Figure 1, the reagent presumably forms a cyclic ethyleneimmonium inter-mediate under the alkaline reaction conditions which then reacts with the starch. When the reaction is finished, acidification is carried out to yield a protonated cat-ionic ammonium ion in a group attached to the starch via an ether linkage. The reagent treatment level is such that the starch remains in granular form so that the product can be recovered by washing, filtration and flash-drying.

An extension of this chemistry to an analogous re-agent containing an activated chlorine in the beta-position to a tertiary amino nitrogen as well as a phosphonic acid group(s) results, under similar reaction conditions, in an amphoteric starch as shown in Figure 2.[3] The reagent itself can be prepared as shown in Figure 3.[4,5] This is not an industrial scale product at this time.

REFERENCES

1. M. W. RUTENBERG, D. SOLAREK, in Starch: Chemistry and Technology, edited by R. L. Whistler, J. N. BeMiller, E. F. Paschall (Academic Press, Orlando, Florida, 1984), Chap. 10, pp. 311–388.

2. C. G. CALDWELL, O. B. WURZBURG, U.S. Patents 2,813,093 (1957), 2,935,436 (1960). Chem. Abstr., 52, 2438 (1958), 54, 16831 (1960).

3. M. M. TESSLER, U.S. Patents 4,243,479 (1981), 4,260,738 (1981). Chem. Abstr., 95, 26916, 117401 (1981).

4. K. MOEDRITZER, R. IRANI, J. Org. Chem., 31, 1603.

5. M. M. TESSLER, U.S. Patent 4,297,299 (1981).

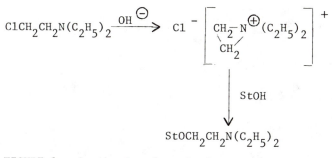

$$ClCH_2CH_2N(C_2H_5)_2 \xrightarrow{OH^{\ominus}} Cl^- \left[\begin{array}{c} CH_2{-}N^{\oplus}(C_2H_5)_2 \\ CH_2 \end{array} \right]^+$$

$$\downarrow StOH$$

$$StOCH_2CH_2N(C_2H_5)_2$$

FIGURE 1   Synthesis of cationic starch

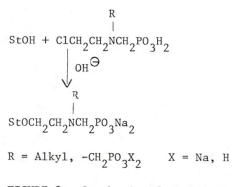

$$StOH + ClCH_2CH_2\overset{\overset{\displaystyle R}{|}}{N}CH_2PO_3H_2$$

$$\downarrow OH^{\ominus}$$

$$StOCH_2CH_2\overset{\overset{\displaystyle R}{|}}{N}CH_2PO_3Na_2$$

R = Alkyl, $-CH_2PO_3X_2$ X = Na, H

FIGURE 2   Synthesis of amphoteric starch

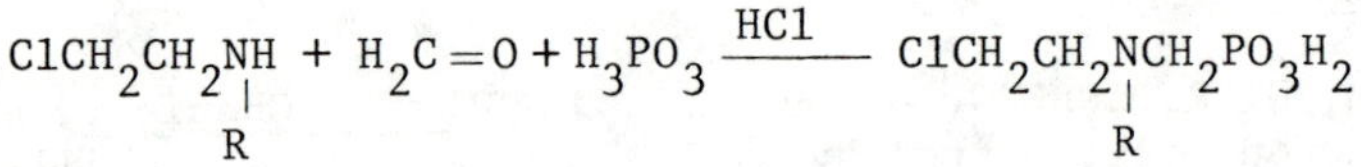

M. W. RUTENBERG

$$ClCH_2CH_2\underset{R}{N}H + H_2C=O + H_3PO_3 \xrightarrow{HCl} ClCH_2CH_2\underset{R}{N}CH_2PO_3H_2$$

FIGURE 3   Synthesis of aminophosphonic acids

NEW APPROACHES TO THE SYNTHESIS OF BRANCHED POLYSACCHARIDE DERIVATIVES FOR STRUCTURE/FUNCTION STUDIES

Manssur Yalpani
Chemical Technology Division
B.C. Research
3650 Wesbrook Mall
Vancouver, B.C. V6S 2L2
Canada

Synthetic methods are discussed for the preparation
of new types of branched polysaccharides from
chitosan, cellulose and galactomannan precursors.
The utility of these techniques for the modification
of polysaccharide properties is exemplified,
providing an indication of their potential value in
the context of systematic structure/function studies.

1. INTRODUCTION

The increasing prominence of polysaccharides in various

industrial applications has highlighted the demand for

chemical methods which facilitate the "tailoring" of

polymer properties, and provided renewed impetus to

investigations of their physical characteristics,

particularly at the molecular level. As a result,

considerable progress has been made in our understanding

of the factors governing primary and higher-order solution

structures of polysaccharides.[1] Some general features
of the rheological behaviour of these polymers and their
interactions with various types of solutes have also been
established.[2]

However, despite these advances, systematic studies of
the polysaccharide structure/function relation have been
severely limited by the absence of suitable methods for
the preparation of model compounds, a prerequisite for
which is the availability of selective, yet facile,
derivatization techniques. This note summarizes some
recent developments in the synthesis of branched model
polysaccharide derivatives.

2. Branched Chitosan Derivatives

One of the most suitable substrates for the preparation of
branched model polysaccharides is chitosan, the 2-amino-2-
deoxy function of which offers a convenient locus for
chemical modifications. Thus, specific attachment of
carbohydrate residues to this position transforms the
intractable, linear polymer into branched, water soluble
derivatives, whose degree of branching, branch length,
type and stereochemistry are amenable to facile
modifications.[3] These conversions can be accomplished
by several methods, including by reductive alkylation,
Schiff's-base formation, and amidation reactions, and
using a variety of carbohydrate types.

The reductive alkylation of chitosan can, for example,
be readily conducted at ambient temperature using
carbonyl-containing mono-, di-, oligo-, or
poly-saccharides and sodium cyanoborohydride, to afford
the corresponding branched derivatives in high yields

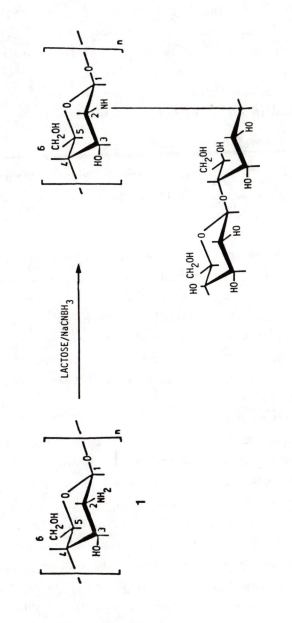

SCHEME 1

(70-100%, depending on the size of the saccharidic residue)[3], as illustrated in Scheme 1 and in Figure 1. The procedure can equally be extended to incorporate keto-sugars, such as fructose (Figure 2, structure 16), and less common types of carbohydrates, such as streptomycin (14), selectively oxidized cyclodextrins (15), etc.[3]

On the other hand, branched derivatives with hydrolytically more labile imine or amide linkages can be obtained in similar fashion by omission of the reducing agent, or by employing sugar acids or lactones, respectively.

The above series of selective modification methods provide a unique opportunity for the preparation of a whole family of closely related polysaccharide derivatives. The distinguishing features of the members of this family may range from relatively subtle structural differences, such as the variation in branch length by a single carbon, to major variations in their properties, such as branch type, net charge, etc.

A different approach to the modification of polysaccharide properties involves derivatizations which employ a combination of saccharidic and non-saccharidic carbonyl reagents. Drastic changes in polymer characteristics can be induced by using an appropriate choice of reagent types and ratios. Thus, reductive alkylation of chitosan with a mixture (1:1.3) of lactose and propionaldehyde, instead of only lactose (product 2), affords a copolymer (Figure 3, 17) which is not only water soluble but also compatible with organic solvents.[3] Similarly, incorporation of low concentrations of the

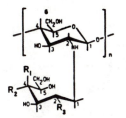

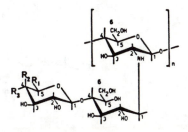

	R₁	R₂	R₃
3	H	OH	OH
4	H	OH	NHCOCH₃
5	H	OH	NH₂
6	OH	H	OH
7	OH	H	NH₂

	R₁	R₂	R₃
2	CH₂OH	OH	H
11	CH₂OH	H	OH
8	CHO	OH	H

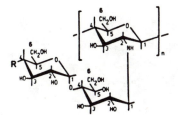

R

9 H

10

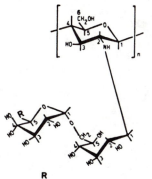

R

12 CH₂OH

13 CHO

FIGURE 1

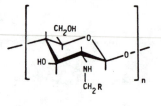

R

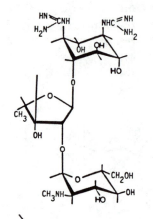

14

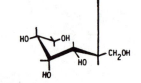

15

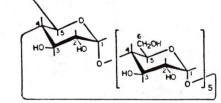

16

FIGURE 2

hydrophilic lactose component into an otherwise completely insoluble salicylidene chitosan derivative affords a water soluble product (18) with improved metal chelating capacity.[4]

3. Other Branched Polysaccharide Derivatives

The synthesis of branched model polysaccharides from starting materials that lack a prominent functional group requires selective activation techniques. A number of such methods have been developed recently. For example, cellulose can be selectively oxidized at C-3 using the acetic anhydride-methyl sulfoxide ($Ac_2O/DMSO$) oxidant system in DMSO-paraformaldehyde, to obtain 3-oxycellulose 19 (scheme 2). The corresponding 2-oxy-cellulose 20 can be obtained using a 6-O-triphenyl methane-cellulose precursor and $Ac_2O/DMSO$. Selective oxidations of the primary hydroxyl function of the galactosyl residues of galactomannans, such as guar gum and locust bean gum, or other galactose-containing polysaccharides can be accomplished by galactose oxidase (E.C. 1.1.3.9) treatment to afford the corresponding C-6 aldehyde derivatives.

The above carbonyl and aldehyde derivatives are amenable to the facile reductive amination reactions, either directly, using aminosugars, such as glucosamine, or via conversion to a primary amine, using ammonium acetate, followed by reductive alkylation with a carbonyl-containing carbohydrate, such as lactose, as exemplified by the branched, water soluble cellulose derivative 21 (Scheme 2) and the guar gum derivative bearing an extended side-chain 23 (Scheme 3).

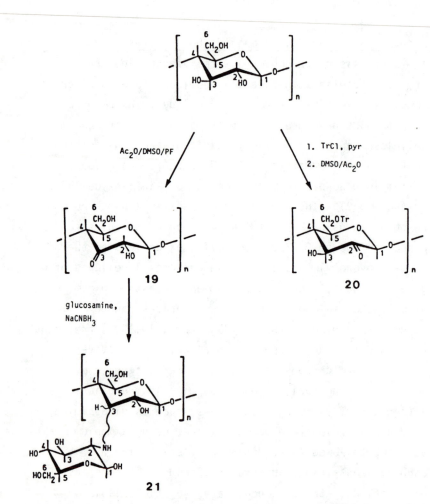

SCHEME 2

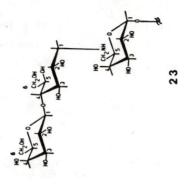

23

1. Galactose Oxidase
2. NH_4OAc / $NaCNBH_3$
3. Lactose / $NaCNBH_3$

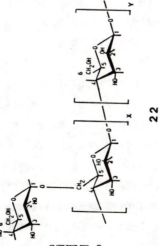

22

SCHEME 3

Carbohydrate branches can also be selectively
introduced at various other polysaccharide sites, as
exemplified by various reducing-end modifications of
dextran and other polymers.[7]

4. CONCLUSION

A particularly attractive aspect of the synthetic
techniques discussed here is the potential of modifying a
wide range of physical properties of polysaccharides, such
as their solubility, hydrophilic-lipophilic balance,
compatibility, and rheology.[6,8] The methods are simple
and efficient and in combination with various selective
activation techniques, can be applied to numerous types of
polysaccharides, offering the prospect of systematic
variations of a host of branch residue parameters. These
structural variations are schematically summarized in
Figure 4, where heavy horizontal lines represent the
polymer backbone, to which cyclic (circles) and acyclic
(indicated by heavy vertical lines, with each carbon
center being represented by a perpendicular line)
carbohydrate residues are attached.

Thus, attachment of appropriate aldehydo-, keto-, or
lactone sugars to the polymer backbone affords derivatives
with varying lengths (from C_3 to C_7 and higher
residues) of short, acyclic branches, for which the
stereochemistry at almost all carbon centers can be
modified by simple choice of a suitable stereoisomer
(Figure 4 A-D). Similarly, medium-sized ($C_{12} - C_{18}$
..residues), or long branches may be introduced using

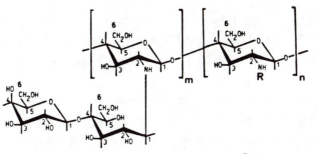

R   17 (CH₂)₂CH₃   18

FIGURE 3

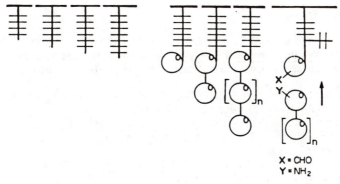

X = CHO
Y = NH₂

A   B   C   D   E   F   G   H

X = NH₂
Y = C=O

m = 1,2,3

I   J   K   L   M

FIGURE 4

di-,oligo- or poly-saccharides (Figure 4 E-G), or by post
modification of preformed branches (derivatives 8 and 13,
Figure 4 H).

The branch type can be varied by introduction of
structural irregularities, using, for example, 2-keto
sugars (eg. derivative 16, Figure 4 I), disaccharides with
glycosidic linkages other than 1-6 (e.g., derivative 2,
Figure 4 J), or by modificaiton of suitably functionalized
branch residues (eg. derivative 5 and 7, Figure 4 K).
Other branch types are exemplified by derivatives 14
(Figure 4 L) and 15 (Figure 4 M).

REFERENCES

1. D.A. REES, E.R. MORRIS, D. THOM, and J.K. MADDEN,
    in The Polysaccharides, edited by G.O. Aspinall
    (Academic Press, New York, 1982), Vol. 1, Ch. 5,
    pp. 195-290.
2. E.R. MORRIS and S.B. ROSS-MURPHY, in Techniques in
    Carbohydrate Metabolism, edited by D.H.
    Northcote; (Elsevier, Amsterdam, 1981), B310.
3. M. YALPANI and L.D. HALL, Macromolecules, 17,
    272-281, 1984.
4. L.D. HALL and M. YALPANI, J. Chem. Soc. Chem.
    Commun., 1153-1154, 1980.
5. M. YALPANI, L.D. HALL. J. DEFAYE, and A. GADELLE,
    Can. J. Chem., 62, 260-262, 1984
6. M. YALPANI and L.D. HALL, J. Polym. Sci., Polym.
    Chem. Edn., 20, 3399-3420, 1982.
7. M YALPANI and D.E. BROOKS, submitted for
    publication.
8. M. YALPANI, L.D. HALL, M.A. TUNG, and D.E.
    BROOKS, Nature, 302, 812-814, 1983.

NEW POLYURONATES FROM NATURAL GLUCANS

A. CESÀRO, F. DELBEN, T.J.PAINTER[1] and S. PAOLETTI
Institute of Chemistry
University of Trieste
I-34127 Trieste, Italy

By addition of finely-powdered $NaNO_2$ to solutions
of natural polysaccharides in 85% w/w orthophospho
ric acid, a series of new polyuronates was prepared,
having Mn of ca. 10^4 Daltons. Characterization of
the products showed that: i) severe depolymerization occurs (depolymerization apparently results
from non-specific oxidation); ii) a very high degree
of substitution was obtained; iii) the products
exhibit the typical polyelectrolytic behavior in
solution; iv) aqueous solutions of the products
show the characteristic circular dichroic spectrum
of the carboxylic group; v) the acidic form of the
samples show different solubility, which depends
on the stereochemistry of the glycosidic bonds, and
hence on the flexibility of the chain; vi) the
products interact with copper(II) ions.

1. INTRODUCTION

Conversion of neutral glycans into ionic polymers is of
great interest for new potential uses of widely distributed
materials. Among various reactions, selective oxidation of
cellulose by $NaNO_2$ in 85% orthophosphoric acid was recently
described.[2] In this reaction, the liberated N_2O_3 generated

307

a foam which was stable for many hours, because of the viscosity of the solution. The smooth and rapid oxidation observed was ascribed to the high surface area within the
foam and the excess pressure within the bubbles. Characterization of new glucuranoglucans prepared by application
of the same methods to various substrates is reported.
Substrates considered were amylose and scleroglucan, a
fungal polysaccharide constituted by a linear chain of
1,3-β-glucopyranose units with single 1,6-β-glucopyranose
residues linked to every third unit of the chain. The aim
was to obtain a series of new (C-6 oxidized) polyuronic
acids, which vary in stereochemistry of the monomeric unit
and of the glicosidic linkage as well as in the extent of
oxidation. Investigation of their solution properties was
carried out to disclose, in addition to the expected ionic
behavior, their capability in metal cation binding and in
gel formation. Other natural polyuronates, like alginate
and pectate , in fact are widely employed in industy as
gelling agents.

Solution properties of polyuronates have been extensively studied in our laboratory during the past years.[3-9]

2. PREPARATION OF C-6 DERIVATIVES

Details about the preparation of C-6 oxidized amylose samples with variable degree of oxidation and the characterization of these products are reported elsewhere.[10] In the
preparation of C-6 oxidized scleroglucan, the reduction
with $NaBH_4$ continued for five hours. The clear yellow cen-
trifugate obtained from the solution at pH 6 was dialyzed
against distilled water. The solution was then adjusted to
pH 3 with HCl, and ethanol was added. The white precipitate

was redissolved in water, dialyzed against water, and finally
freeze-dried. The solution of C-6 oxidized cellulose, after
reduction by $NaBH_4$ (12 hours), was neutralized with acetic
acid, filtered, dialyzed against 1N acetic acid, water, and
ethanol, in the order. The gel formed was squeezed onto a
filter and washed with ethanol and with diethyl ether.
The solid was dried in a vacuum over at 40°C over $CaCl_2$;
the yields were 86 up to 93%.

3. RESULTS AND DISCUSSION

3.1 Non-selective oxidation and depolymerization

Aqueous solutions of the initial oxidized products turned
yellow and showed a decrease in viscosity, when made alkaline. Presence of keto-groups in these products was evident from their reactivity with phenylhydrazine and caused
the degradation by β-elimination. Reduction of keto groups
with $NaBH_4$ proceeded rapidly, as shown by the disappearance
of the yellow color, and with considerable stereoselecti-
vity, because after hydrolysis only D-glucuronic acid was
detected among acidic products. Because of the strong acidic
conditions, it is expected that hydrolysis occurs simultaneously with oxidation. However, the most severe form of
depolymerization is likely to arise from non-specific
oxidation. In fact, reoxidation of amylose without prior
reduction with $NaBH_4$ produced a material highly depolymerized that was not retained by the dialysis membrane.
On the other hand , it is interesting to report that reoxidation after borohydride reduction gave a C-6 oxidized
amylose in comparatively good yield (about 86%) and without
further depolymerization. Therefore, depolymerization can

be controlled by carrying out the oxidation in steps, with
borohydride reduction after each. However, under alkaline
conditions of borohydride treatment, depolymerization by
β-elimination would occur as a competitive reaction. The
rate of β-elimination could be greatly decreased simply by
increasing the concentration of $NaBH_4$ in the solution[11]
from the value of 2% w/v, used in the present work, up to
20% w/v.

3.2 Physico-chemical characterization of the products

C-6 oxidized amylose and scleroglucan were freely soluble
in water in both their Na^+ and H^+ forms. On the contrary,
C-6 oxidized cellulose was soluble as Na^+ salt but formed
bright gels at low pH values. All the products were otherwise typical glycuronoglycans, giving insoluble salt with
Ca^{2+}, Sr^{2+}, Ba^{2+} and heavy-metal cations, and with cetylpyridinium and cetylmethylammonium ions.

Number-average molecular weights were determined from
membrane osmometry measurements and resulted 10^4 to 1.5×10^4
Dalton for all the derivatives. Intrinsic viscosities of
the salt forms determined at 25°C were found to depend on
both polymer species and molecular weight.

Potentiometric titration curves of oxidized cellulose
and amylose in water exhibit the typical trend of a poly-
acid bearing one type of ionizable groups (Figure 1).
On the contrary, the potentiometric curve of C-6 oxidized
scleroglucan shows two distinct inflections, ascribable to
the presence of two series of carboxyl groups with different
strength. Although the chemical structure of oxidized
scleroglucan could allow a qualitative explanation for this
effect, other hypothesis of conformational nature cannot be

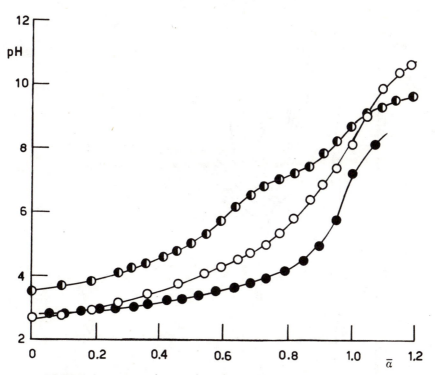

FIGURE 1 Potentiometric titration curves of C-6 oxidized polysaccharides in water at 25°C. C-6 oxidized cellulose, ● ; amylose, ○ ; scleroglucan, ◐ . C-6 oxidized cellulose experiment refers to back-titration.

excluded.

New circular dichroic bands associated with n → π* and π → π* electronic transitions of carboxylate groups appeared in the region from 190 to 250 nm. This bands undergo a strong change in intensity, with a clear isodichroic point, when pH of solution was changed from neutral to acidic values. As an example, titration of scleroglucan is reported in Figure 2. Chiroptical measurements were carried out on a series of C-6 oxidized amylose samples having various

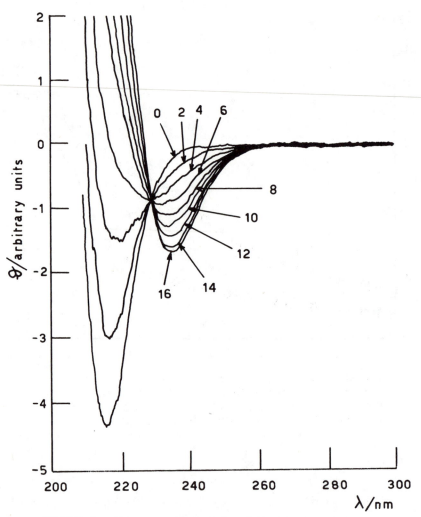

FIGURE 2   CD spectra of C-6 oxidized scleroglucan in aqueous 0.05 M NaClO$_4$ at 25°C as a function of the stated amounts (µL) of 1.23 N HClO$_4$ added to 2 mL solution of polymer in Na$^+$ salt form.

degrees of oxidation. By using these optical rotation data and that of maltose from literature a specific rotation of 280° was calculated for the 1,4-α-$\underline{\underline{D}}$-glucuronate residue.

Formation of blue complex by C-6 oxidized amylose samples with triiodide and iodine was detected by UV-visible
absorption and CD spectroscopy. A large shift of the λ max
to smaller values with respect to that of parent amylose
was found consistent with the depolymerization. However,
the retention of chirality of CD bands suggests that the
left-handed helical conformation is still the most stable
one, despite the change of neutral primary alcoholic groups
to carboxylate groups.

CD spectra of aqueous solutions of the samples in the
presence of copper ions show the appearance of a new band
centered around 250 nm. This band, already found in other
polyuronate $-Cu^{2+}$ systems, is due to a charge $-$transfer
from the COO^- to the Cu^{2+} ions and reflects an intimate
donor-acceptor interaction. The intensity of this band
varies among the samples investigated also because of its
sensitivity to stereochemical environment.

4. <u>CONCLUSIONS</u>

In its present form, the method will certainly not give products that are both highly oxidized and high in molecular
weight. Since the oxidizing ability of the mixture depends
on the stability of the foam generated by N_2O_3 and hence
on the viscosity of the initial solution in orthophosphoric
acid, a way is needed to increase that viscosity. This is
particularly important for amylose, whose solutions exhibit
exceptionally low viscosities. An inert polymer might be
used. The simpler device of increasing the concentration of
the orthophosphoric acid solution above 85% w/w may lead
to some phosphorilation of the polysaccharide. The addition
of buffer salts might be used to control acid-catalised

hydrolysis during the reaction, but fundamental studies of the oxidation of primary and secondary alcohols by N_2O_3 are needed, to predict, and, if possible, control the extent of non-specific oxidation. On the credit side, it should be noted that the starting material is largely present in nature, and that the oxidized products appear to have very different solution properties. Moreover, both the excess of orthophosphoric acid and the ether, used to extract it, could be recovered by distillation, and that the only by-product of the oxidation, sodium phosphate, is non-toxic.

5. ACKNOWLEDGEMENTS

This work was sponsored by Consiglio Nazionale delle Ricerche (Rome), Progetto Finalizzato "Chimica fine e secondaria".

REFERENCES

1. On leave of absence from the Institute of Marine Biochemistry, University of Trondheim, N-7034 Trondheim NTH (Norway).
2. T. J. PAINTER, Carbohydr.Res., 55, 95-103 (1977).
3. A. CESARO, A. CIANA, F. DELBEN, G. MANZINI and S. PAOLETTI, Biopolymers, 21, 431-449 (1982).
4. F. DELBEN, A. CESARO, S. PAOLETTI and V. CRESCENZI, Carbohydr.Res., 100, C46-50 (1982).
5. A. CESARO, F. DELBEN and S. PAOLETTI, Thermodynamics Properties of the Polyuronates - Ca^{2+} Ions Interaction in Aqueous Solutions, Thermal Analysis, edited by B. Miller (J.Wiley & Sons, Chichester, 1982), pp.815-821.
6. A. CESARO, S. PAOLETTI, F. DELBEN, V. CRESCENZI, R. RIZZO and M. DENTINI, Gazz.Chim.Ital., 112, 115-121 (1982).
7. S. PAOLETTI, A. CESARO and F. DELBEN, Carbohydr.Res., 123, 173-178 (1983).
8. G. MANZINI, A. CESARO, F. DELBEN, S. PAOLETTI and E. REISENHOFER, Bioelectrochem.and Bioenergetics, 12, 000 (1984), in press.

9. E. REISENHOFER, A. CESARO, F. DELBEN, G. MANZINI and
   S. PAOLETTI, Bioelectrochem. and Bioenergetics, 12,
   000 (1984), in press.
10. T. J. PAINTER, A. CESARO, F. DELBEN and S. PAOLETTI,
    Carbohydr.Res., in press.
11. T. J. PAINTER and B. LARSEN, Acta Chem.Scand., 27,
    1957-1962 (1973).

DISCUSSION OF DEREK HORTON'S PAPER

M. HEIDELBERGER
Department of Pathology
New York University School of Medicine
New York, N.Y. 10016

I find Derek Horton's relatively mild method
of oxidizing cellulose at carbon 6 most interesting
for the following reasons: In 1942, on reading
W.O. Kenyon's papers [1,2] in the Journal of the
American Chemical Society, I realized at once that
he and his group at Eastman Kodak had, by oxidizing
cellulose with nitric oxide into a water-soluble
polysaccharide with 16-20% carboxyl groups at C6,
converted it into a substance greatly resembling
the type-specific capsular polysaccharide of type
8 pneumococcus:[3]

$$\rightarrow 4)\,\text{DglcA}\,\beta(1\rightarrow 4)\,\text{Dglc}\,\beta(1\rightarrow 4)\,\text{Dglc}\,(1\rightarrow 4)\,\text{Dglc}\,(1\rightarrow]_n$$

Dr. Kenyon kindly sent me a sample, some of which
I dissolved with the help of sodium bicarbonate and
added to antipneumococcal type 8 horse serum, get-
ting an immediate precipitate [4]. Was I excited!

It had long been known [5] that type 3
pneumococcal polysaccharide, which is a polymer of
cellobiouronic acid [6], cross-reacted with type 8
antiserum, as also did the type 8 substance in
antipneumococcal type 3 serum. Quantitative analysis
of the precipitates and supernatants showed that

the same fraction of antibody was precipitated in
the cross-reactions by the oxidized cellulose and
the type 3 and type 8 polysaccharides [4]. We even
used the solution of oxidized cotton as a vaccine
in an attempt to protect mice against type 8
pneumococci but failed, perhaps because of
depolymerization caused by the rather drastic
method of oxidation. Possibly Derek's product would
have been effective.

There was, however, an immediate pratical use
for Kenyon's oxidized cotton, as I discovered in
talking about the cross-reactions in the faculty
lunchroom at Columbia's College of Physicians and
Surgeons where I was working at the time. Someone
said: "Our surgeons are using pads of that stuff
as a hemostatic that can even be left in the site
of operation without danger because it eventually
dissolves." How long that took was unknown, but
after surgeons of the Presbyterian Hospital began
sending me samples of patient's blood or urine I
could tell them almost to the hour when the last
traces of the pad had disappeared.

REFERENCES

1. E.C. YACKEL & W.O. KENYON, J. Am. Chem. Soc.,
    64, 121- 127, (1942)
2. C.C. UNRUH & W.O. KENYON, J. Am. Chem. Soc.,
    64, 127-131, (1942)
3. J.K.N. JONES & M.B. PERRY, J. Am. Chem. Soc.,
    79, 2787-2793, (1957)
4. M. HEIDELBERGER & G.L. HOBBY, Proc. Nat. Acad.
    Sci. (U.S.), 28, 516-518, (1942)
5. J.Y. SUGG, E.L. GASPARI, W.L. FLEMING, & J.M.
    NEILL, J. Exp. Med., 47, 917-931; (1928);

G. COOPER, M. EDWARDS, & C. ROSENSTEIN, J.
Exp. Med. 49, 461-474, (1929)

6. R.D. HOTCHKISS & W.F. GOEBEL, J. Biol. Chem.,
    121, 195-203, (1937); R.E. REEVES & W.F. GOEBEL,
    J. Biol. Chem., 139, 511-519, (1941)

STEREOREGULAR ACYCLIC POLYMERS
FROM GLYCOL-SPLIT POLYSACCHARIDES

B. CASU

Istituto di Chimica e Biochimica "G. Ronzoni"
via G. Colombo, 81 - 20133 Milano, Italy

Splitting of 1,2-diols (with periodates, or lead tetraaceta_ te) is currently used in structural analysis of carbohydra_ tes and their polymers, mainly for detecting and quantita_ ting unsubstituted vicinal hydroxyls.[1]   Advantage is also taken of the reactive aldehydo groups formed at the sites of splitting, for functionalizing polysaccharides, most com_ monly only at the level of a few residues.[2]   However, split_ ting reactions have not been deliberately used for modify_ ing the chain geometry and flexibility, and hence the macro_ molecular properties, of polysaccharides.   This modifica_ tion can be especially dramatic for 1→4-linked polysacchari_ des, where each "opened" pyranose ring is expected to act as a flexible joint along the polymer chain.[3,4]

In the framework of a systematic investigation on the influence of splitting reactions and further derivatization of products of splitting, on the physico-chemical, conforma_ tional and rheological properties of polysaccharides, all the hexopyranose rings of 1→4-linked polysaccharides have been oxidized with periodate under conditions for minimizing depolymerization, and the resulting polydialdehydes either further oxidized (with chlorite) to obtain poly-dicarboxy lates, or reduced with borohydride to obtain polyalcohols, as illustrated in Figure 1 for glucans.   Polyalcohols have been also converted into the corresponding polyacetates. The NMR spectra indicated high degrees of conversion and structural homogeneity of the products.

Products from glucans were the polyelectrolytes 2,3-di carboxy-cellulose (DCC) and -amylose (DCA),[5] and the polyal cohols RDC and RDA.[6]   DCC and DCA bind calcium and magnes ium,[7] a property at the basis of the previously reported good performance as builders in detergents,[8] (and, probably, of antiviral properties[9]) of these and similar carboxylated polysaccharide derivatives.   The poly-alcohols and the cor responding polyacetates are essentially new polymers, with remarkable tendence to aggregate in solution, and crystalli ze.[6]   A X-ray fiber pattern compatible with a two-fold he lix was obtained from the acetate of the cellulose derivati ve (RDC-Ac).[10]   Structurally homogeneous RD-derivatives we re obtained also from α and β D-galactans, and poly-α-D-ga lacturonate.[11]   As indicated by the NMR spectra, products retained the configuration at C-1 and C-4 of the original polysaccharides. (Formulas for RD derivatives of glucans and galactans are shown in Figure 2.) Solution properties potentially associated with stereoregularity of the polymers are being further investigated.

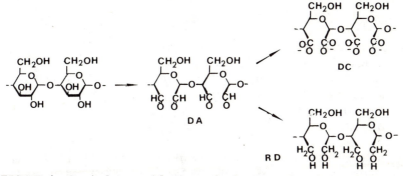

FIGURE 1. Periodate oxidation of glucans, to give polydialdehydes (DA), and further oxidation to polydicarboxylates
(DC), or reduction to polyalcohols (RD).[4]

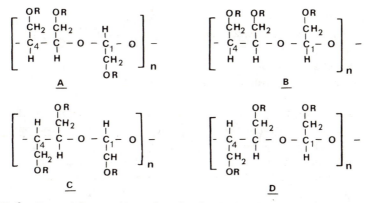

FIGURE 2. Repeating units of polyalcohols (R=H) and polyacetates (R=Ac) obtained from 1→4-linked β-D-glucan (A), α-D-
glucan (B), β-D-galactan (C), and α-D-galactan (D).

Glycol splitting can be used also for modifying side-
chains of branched polysaccharides, as reported for scleroglucan, whose "dicarboxylated" derivatives have interesting
solution and gel-forming properties.[12,13]   Xanthan is being
systematically modified by splitting of side-chain residues,
with or without removal of acetate groups.[14]

REFERENCES

1. A. S. Perlin, Glycol-Cleavage Oxidation, in The Carbo _
   hydrates, edited by W. Pigman, D. Horton, and J. D. Wan_
   der, Academic Press, New York, Vol. IB, 1980, pp.1167.
2. J. F. Kennedy, Adv. Carbohydr. Chem. Biochem., 29,
   306-405 (1974).
3. O. Smidsrød and T. Painter, Carbohydr. Res., 26, 125-
   132 (1973).
4. B. Casu, S. V. Meille, A. Naggi, G. Torri, G. Zoppetti,
   and G. Allegra, Carbohydr. Polym., 2, 283-287 (1982).
5. B. Casu, U. Gennaro, S. V. Meille, M. Morrone, A. Naggi,
   M. S. Occhipinti, and G. Torri, Int. J. Biol. Macrom.,
   8, 89-92 (1984).
6. B. Casu, G. Torri, A. Naggi, A. Allegra, S. V. Meille,A.
   Cosani, and M. Terbojevich, Macromolecules, submitted.
7. V. Crescenzi, M. Dentini, C. Meoli, B. Casu, A. Naggi,
   and G. Torri, Int. J. Biol. Macrom., 6, 142-144 (1984).
8. U.S. Patent 3,629,121 (1971).
9. P. Claes, A. Billiau, E. De Clerg, J. Desmyter, E.
   Schonne, H. Vanderhaeghe, and P. De Somer, J. Virology,
   5, 313-320 (1970).
10. S. V. Meille, G. Allegra, A. Naggi, G. Torri, and B.
    Casu, Abstracts XIIth Intern. Carbohydrate Symposium,
    Utrecht, July 1984, p. 525.
11. B. Casu, A. Naggi, and G. Torri, unpublished results.
12. V. Crescenzi, A. Gamini, G. Paradossi, and G. Torri,
    Carbohydr. Polymers, 3, 273-286 (1983).
13. A. Gamini, V. Crescenzi, and R. Abruzzese, Carbohydr.
    Polymers, 4 (1984), in press.
14. B. Casu, P. Giovannetti, A. Naggi, and G. Torri, unpu_
    blished results.

   This work is part of a collaborative research supported
by the Italian Research Council (CNR), Target-Oriented Pro_
ject on Fine and Secondary Chemistry, involving the Polys_
accharide Laboratory of the Ronzoni Institute, the Depart_
ment of Chemistry of the University of Rome, the Department
of Chemistry of the Polytechnic of Milan, and the CNR Labo_
ratory of Biopolymers of the University of Padua.

POLYSACCHARIDES AND CONJUGATED POLYSACCHARIDES AS HUMAN VACCINES

HAROLD J. JENNINGS
Division of Biological Sciences
National Research Council of Canada
Ottawa, Ontario, Canada K1A OR6

Capsular polysaccharides have assumed an important
role as vaccines against disease caused in humans by
bacteria. The concept of using purified
polysaccharides devoid of the accompanying bacterial
mass is technically elegant as they are non-toxic,
immunogenic, and can be chemically and physically
defined. The concept is also capable of extension
into other areas of immunoprophylaxis. However three
problems have been identified which interfere with the
complete adoption of this concept. These are the
problem of the multiplicity of polysaccharides
required to give complete protection, the poor
immunogenicity of polysaccharides in infants, and the
poor immunogenicity of the group B meningococcal
polysaccharide in all humans. Solutions to these
problems are proposed which involve the synthesis of a
new-generation of polysaccharide- or oligosaccharide-
protein conjugates. These conjugates can be formed by
random or selective activation of the polysaccharide
prior to its conjugation and a critical analysis of
these two procedures is given.

1. INTRODUCTION

Vaccination has proved to be one of the most useful
scientific developments in the control and eradication of
human disease. Early vaccines were based on whole-cell

preparations, and the concept of using pure chemically and
physically definable polysaccharides as vaccines against
disease caused by bacteria is technically elegant[1,2,3].
This concept stems from early research on Streptococcus
pneumoniae when their capsular polysaccharides were
identified as powerful type specific antigens in both
animal and human responses to these organisms[4,5]. Two
major polysaccharide vaccines against human disease caused
by bacteria are in current use. These vaccines have proved
to be effective, are non-toxic, and have been administered
to in excess of approximately one hundred million people.
One is a divalent (Groups A and C polysaccharides) vaccine
against meningococcal meningitis and the other is a
multivalent vaccine against pneumococcal infections
consisting of 14 different type-specific capsular
polysaccharides[1,2,3]. The concept of using capsular
polysaccharides as human vaccines is obviously capable of
extension to disease caused by other pathogenic bacteria,
H. influenzae, Group B Streptococcus, S. aureus,
Klebsiella, and Pseudomonas; however, their use will depend
on their clinical importance and the presence of meaningful
epidemiological studies.

Despite the success obtained with the use of
polysaccharides as human vaccines, a number of problems
associated with the further development of this concept
have been encountered, which have stimulated much recent
research. These problems include the multiplicity of
polysaccharides required to give adequate protection
against some bacteria, the poor immunogenicity of
polysaccharides in infants, the section of the population
most susceptible to bacterial meningitis, and the poor
immunogenicity of the Group B meningococcal polysaccharide

in all humans. The purpose of this lecture is to address
these problems and to discuss strategies to surmount them.

1.1. Multivalent Polysaccharide Vaccines

Because of the diversity of pneumococcal-capsular types (84
have so far been identified) the intent behind the
formulation of the 14 valent pneumococcal vaccine was to
limit the number of polysaccharides while maintaining the
maximum effective coverage. The choice of the 14
polysaccharides (Table 1), estimated to be involved in
70-80% of all pneumococcal bacteremic infections was based
largely on the frequency of occurrence of the different
serotypes in disease isolates[3]. Recent studies have
indicated that an increased level of protection
(approximately 90%) can be obtained using 23
polysaccharides (Table 1) and the formulation of a new
vaccine is under serious consideration[6]. The increased
coverage is largely obtained by the inclusion of type-
specific polysaccharides from less frequently isolated
pneumococci (Table 1), although other important factors
have also influenced these additions.

One property of the pneumococcal polysaccharides that
appeared favorable for the limitation of their number in
the vaccine was their extensive serological cross-
reactivity in animals. This cross-reactivity is based on
structural homology within the polysaccharides and is
exemplified in the Danish serotyping system which
designates capsular types within groups based on this
cross-reactivity (Table 1). However, more recent studies[6]
have demonstrated that some of the polysaccharides do
not provide the same degree of cross-reactivity in humans,

TABLE 1   Types and cross-reactive types within
           pneumococcal groups in the current 14 valent and
           proposed 23 valent polysaccharide vaccine[a].

Group	Cross-reactive types
1*+	None
2*+	None
3*+	None
4*+	None
5*+	None
6	6A*, 6B+
7	7F*+, 7A, 7B, 7C
8*+	None
9	9A, 9L, 9N*+, 9V+
10	10F, 10A
11	11F, 11A, 11B, 11C
12	12F*+, 12A
14*+	None
15	15F, 15A, 15b+, 15C
17	17F+, 17A
18	18F, 18A, 18B, 18C*+
19	19F*+, 19A+, 19B, 19C
20+	None
22	22F+, 22A
23	23F*+, 23A, 23B
25	25F*, 25A
33	33F+, 33A, 33B, 33C

[a]Taken from reference 6.
*Composition of current 14 valent vaccine.
+Composition of the proposed 23 valent vaccine.

the response being largely type-specific. Therefore the
inclusion of additional polysaccharides to the modified
vaccine was proposed to counteract this deficiency. For
example, although the types 19A[7] and 19F[2] polysaccharides
differ structurally by only one linkage (Figure 1) and are
closely immunologically related in animals, the type 19F
polysaccharide, currently used in the 14 valent vaccine,
provides inadequate levels of protective antibodies in

$$\rightarrow 4)\text{-}\beta\text{-}\underline{\text{D}}\text{-Man}\underline{\text{p}}\text{NAc-}(1\rightarrow 4)\text{-}\alpha\text{-}\underline{\text{D}}\text{-Glc}\underline{\text{p}}\text{-}(1\rightarrow 3)\text{-}\alpha\text{-}\underline{\text{L}}\text{-Rha}\underline{\text{p}}\,(1\text{-O-P-O-}$$

$$\overset{\overset{\textstyle O}{\parallel}}{\underset{\underset{\textstyle OH}{\mid}}{}}$$

Type 19A

$$\rightarrow 4)\text{-}\beta\text{-}\underline{\text{D}}\text{-Man}\underline{\text{p}}\text{NAc-}(1\rightarrow 4)\text{-}\alpha\text{-}\underline{\text{D}}\text{-Glc}\underline{\text{p}}\text{-}(1\rightarrow 2)\text{-}\alpha\text{-}\underline{\text{L}}\text{-Rha}\underline{\text{p}}\,(1\text{-O-P-O-}$$

$$\overset{\overset{\textstyle O}{\parallel}}{\underset{\underset{\textstyle OH}{\mid}}{}}$$

Type 19F

FIGURE 1   Repeating units of pneumococcal polysaccharides
            types 19A and 19F.

humans against the type 19A organisms[6]. Thus the inclusion
of the type 19A polysaccharide, in addition to the type 19F
polysaccharide, to the proposed 23 valent vaccine is
advocated (Table 1). A similar argument was used for the
proposed inclusion of the type 9V polysaccharide as well as
the 9N polysaccharide in the 23 valent vaccine.

Although the same serological limitations are not
exhibited by the types 6A and 6B polysaccharides, both
being highly cross-reactive in humans, the choice of type
6B rather than the more frequently encountered 6A
polysaccharide, is based on its relatively greater
stability[8]. This is important in terms of storage of the
vaccine because the immunogenicity of polysaccharides is
dependent on their large molecular size[2]. The structures
of the types 6A and 6B pneumococcal polysaccharides are
shown in Figure 2 and although they differ only in the
position of linkage of their α-L-rhamnopyranosyl residues
to D-ribitol, this structural feature is critical to their

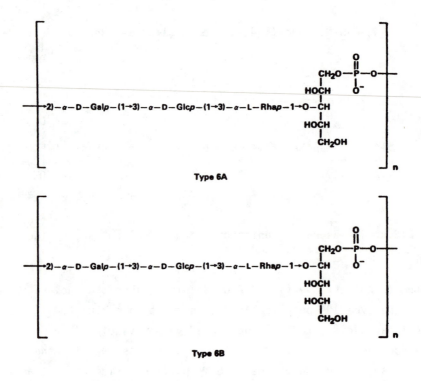

FIGURE 2   Repeating units of the pneumococcal
            polysaccharides types 6A and 6B.

relative stability.   Zon et al.[8] have shown in ^{31}P-nmr
studies on the types 6A and 6B polysaccharides that the
greater instability of the phosphodiester linkages of the
type 6A polysaccharide is due to a hydrolytic mechanism
involving the neighboring group participation of the 4-OH
group of D-ribitol. An equivalent mechanism is not
possible in the case of the type 6B polysaccharide because
the 4-OH groups of its D-ribitol residues are linked
glycosidically to α-L-rhamnopyranosyl residues.
    Problems associated with the assay and control of

multivalent vaccines increase in proportion to the number of polysaccharides used. This fact together with a reluctance to include in the vaccine all the pneumococcal polysaccharides necessary to obtain complete protection, has attracted more recent attention to the possibility of formulating an alternate more simple vaccine based on the pneumococcal subcapsular carbohydrate antigen (C-substance). The repeating unit of C-substance is shown in Figure 3 and has been shown to be common to some different types of pneumococci[9] and is probably common to them all. Recent serological studies have shown that this approach has some merit in that it has been demonstrated by Briles and coworkers[10] that both monoclonal and polyclonal antibodies to phosphocholine (a substituent of C-substance) were protective against a lethal challenge in mice with some encapsulated pneumococci. This observation is also confirmed for some pneumococcal types by Szu and coworkers[11] who also observed that antibodies specific for the complete haptenic structure of C-substance were more effective in mouse protection studies than those specific for phosphocholine alone. Although C-substance is isolated from pneumococci in a small molecular size, poorly

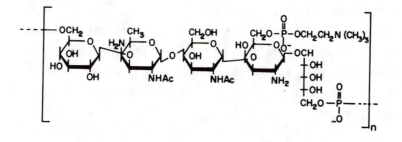

FIGURE 3   Repeating unit of the pneumococcal group antigen (C-substance).

immunogenic form, one strategy to surmount this deficiency
would be to conjugate it to a carrier protein by methods
similar to those described in the following text.

1.2. Poor Immunological Response of Polysaccharides in Infants

As previously described, capsular polysaccharides as human
vaccines are only partially successful due to the fact that
they are poor immunogens in infants[1,2]. The immunological
basis of this phenomenon is the inability of infants to
generate a mature and amnestic response, involving the
production of IgG antibodies, to polysaccharide antigens.
A promising area of research to overcome these deficiencies
is the design of a new generation of semi-synthetic
vaccines based on the conjugation (covalent linkage) of the
polysaccharides to protein carriers[12,13,14]. This
structural modification stimulates the production of
polysaccharide-specific IgG antibodies in mice, a property
not shared by the pure polysaccharide. From the
immunological properties of these conjugates in mice one
can predict that conjugate vaccines have obvious potential
in the formulation of future infant vaccines against
bacterial meningitis.

The feasibility of this approach is well established.
Fifty years ago Goebel and Avery[15] coupled the type III
pneumococcal polysaccharide to horse serum-globulin by the
diazotization of p-aminobenzyl ether substituents on the
polysaccharide, and demonstrated that this polysaccharide
conjugate was able to produce polysaccharide-specific
antibody in rabbits unresponsive to the polysaccharide
alone[16]. Similar experiments were also carried out by

other workers[2]. The coupling procedures employed in this
early work however were far too drastic to be utilized on
most of the highly sensitive polysaccharides currently used
as human vaccines; and resulted in complete random coupling
of the two molecules with the incorporation of highly
undesirable structural features into the conjugates.
Recently more comprehensive studies on polysaccharide-
protein conjugates directed specifically to their use as
human vaccines have now been reported[12,13,14]. The
development of mild, simple and efficient coupling
procedures has resulted in the formation of linkages
containing more innocuous and acceptable structural
features.

　　Two approaches to the synthesis of these conjugates
have been described which involve either random or
selective activation of the polysaccharide prior to its
conjugation to protein. In effect the choice of the method
employed was largely dictated by the molecular size of the
polysaccharide; the larger molecular size polysaccharide of
necessity requiring random activation in contrast to the
facile end-group activation made possible by the use of
polysaccharides of smaller molecular size. The use of
large molecular size polysaccharides in conjugates probably
originated because of the deliberate development of
polysaccharides in this form for the purpose of the pure
polysaccharide vaccines. This strategy was based on the
knowledge that pure polysaccharides in this form function
as better immunogens[2]. Whether this remains true when
polysaccharides are coupled to large immunogenic proteins
seems doubtful although no definitive evidence to resolve
this issue is available at this time.

　　The outcome of this issue will however, probably be

important to the development of polysaccharide-protein
conjugates, because, while random activation of
polysaccharides interferes with their structural integrity
resulting in ill-defined conjugates, the specific
activation of polysaccharide end-group residues leads to
more readily definable conjugates with a high degree of
structural integrity. Also, the reagents used in the
random activation techniques are also more likely to result
in structural modification of the carrier protein. This
could have important implications should one adopt the
strategy of using the carrier protein as a functional
determinant in the vaccine. This would be a likely
strategy in the formation of conjugate vaccines and could
be realized by using as a carrier, either modified
bacterial toxins (e.g. tetanus toxoid and diptheria toxoid)
already in current use as infant vaccines, or other
bacterial surface proteins.

Some examples of conjugation using the random
activation technique are as follows. The large molecular
size H. influenzae type b polysaccharide was conjugated to
a number of proteins by Schneerson and coworkers[12] by
activating the polysaccharide with cyanogen bromide and
functionalizing the protein with an adipic dihydrazide
spacer. The coupling occurs mainly through the formation
of N-substituted isourea bonds but both the polysaccharide
and protein retain unnecessary substituents i.e. carbonate
and underivatized spacer respectively. Large molecular
size group C polysaccharide from N. meningitidis was also
successfully conjugated to tetanus toxoid by Beuvery and
coworkers[14] directly by the use of 1-(3-dimethylamino-
propyl)-3-ethylcarbodiimide hydrochloride (DEC). This
method introduces active O-acylisourea groups at many of

the polysaccharides carboxyl groups, some of which in the presence of protein, form the required amide bonds of the conjugate. However, the introduction of O-acylisourea groups can cause drastic structural changes in the polysaccharide as a result of internal lactonization[17] (Figure 4) and the quenching of the remainder with ethanolamine introduces permanent unwanted substituents

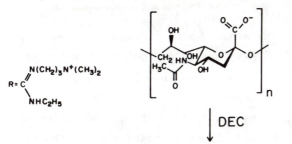

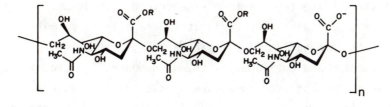

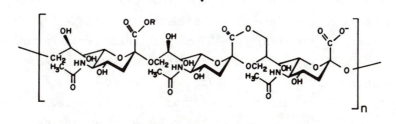

FIGURE 4   Treatment of group C meningococcal polysaccharide with DEC depicting the formation of intramolecular lactones from its O-acylisourea derivative.

into the polysaccharide[14]. Some structural modification of
the carrier protein is also highly likely using DEC as
reagent.

   To develop a more specific approach to the conjugation
of meningococcal polysaccharides to protein they were
monofunctionalized prior to their conjugation to tetanus
toxoid[13]. Controlled periodate oxidation introduced unique
terminal free aldehyde groups into the groups B and C
polysaccharides and in the group A polysaccharide modified
by reduction of its terminal 2-acetamido-2-deoxy-$\underline{D}$-mannose
residue (Figure 5). These monovalent polysaccharides were
then coupled to tetanus toxoid by reductive amination. The
advantage of this method over those previously described,
is that it minimizes the possibility of polysaccharide or
protein modification, and also the possibility of cross-
linking. Like the previously described methods the groups
A and C polysaccharide conjugates were able to stimulate in
mice the production of polysaccharide-specific IgG
antibodies indicative of a "thymus dependent" antigen.
This is demonstrated in Figure 6 which depicts assays of
the antisera from mice previously injected with the group C
polysaccharide conjugate, for antibody (IgG) specific for
the whole conjugate and for the group C polysaccharide
using an enzyme-linked immunosorbent assay (ELISA)[13]. In
contrast to the pure group C polysaccharide, which elicited
a very weak immune response in mice, its tetanus toxoid
conjugate was able to stimulate high levels of
polysaccharide-specific IgG antibodies in the mice.

   No comparative data on the relative immunological
performance of conjugates made by random or selective
activation of the polysaccharides is currently available.
However, should the latter approach produce as effective

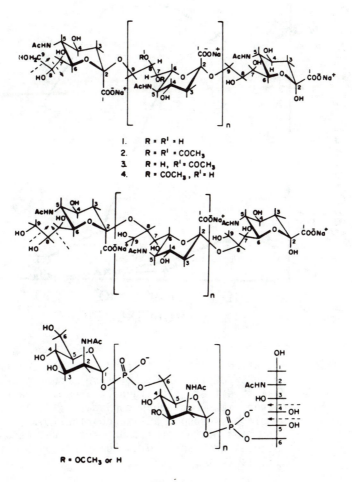

FIGURE 5   Structures of meningococcal groups C (upper), B
            (middle) and end-group reduced A (lower)
            capsular polysaccharides depicting their
            periodate sensitive terminal residues.

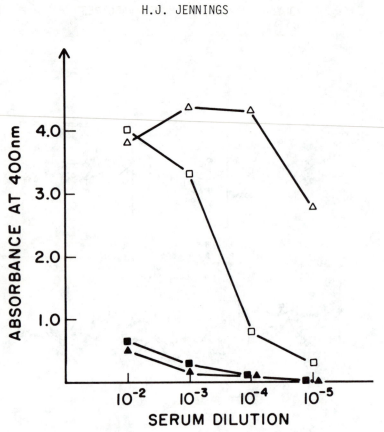

FIGURE 6   Titration of the anti-group C polysaccharide-
            tetanus toxoid conjugate mouse serum in ELISA
            against different antigens. The wells were
            coated with homologous conjugate (Δ), and the
            homologous group C polysaccharide (□).
            Titration of the preimmune sera with the same
            antigens are marked with identical but solid
            symbols (▲,■).

immunogens as the former, then, because it produces more
readily definable conjugates, selective activation of the
polysaccharide should prove to be the method of choice in
the synthesis of conjugate vaccines.

1.3. Poor Immunogenicity of the Group B Meningococcal Polysaccharide in Humans

The group B meningococcal polysaccharide is only poorly immunogenic in humans and animals in both its native[18] and conjugated[13] forms. This is due to immune tolerance, which is probably based on common structural features (repeating α 2→8-linked sialic acid residues) identified in the group B polysaccharide[2] and in the gangliosides of human and animal fetal brain tissue[19]. It has been demonstrated[19] that these gangliosides will bind to horse antibodies (IgM) specific for the group B meningococcal polysaccharide and that they will also inhibit the serological reaction between the group B polysaccharide and the above homologous antibody. Therefore another approach to making a group B meningococcal vaccine based on carbohydrate structures would be to use sub-capsular lipopolysaccharide (LPS) serotype antigens, of which might have been identified in group B strains of meningococci[20]. Due to their highly toxic nature the LPS cannot be used as vaccines. However, mild hydrolysis of the LPS yielded core oligosaccharides (R-type) which contain the major serological determinants of the LPS[21]. The oligosaccharides have a molecular size of 1200 to 1600 daltons and although they have substantial areas of structural similarity they differ sufficiently in structure to account for the LPS type-specificity. The structure of one of the major serotype oligosaccharides (L3), minus its O-ethanolamine phosphate residues is shown in Figure 7 and in gross structural detail is representative of them all[22].

In order to form effective immunogens from these haptenic structures, they were conjugated to carrier proteins[23]. A simple direct coupling of the βD-Gal<u>p</u>(1→4)βD-Gl<u>c</u>pNAc(1→3)βD-Gal<u>p</u>(1→4)βD-Gl<u>c</u>p(1→4)L<u>α</u>D-Hep<u>p</u>(1→5)KDO

                                                        3
                                                        ↑
                                                        1

L<u>α</u>D-Hep<u>p</u>

             2
             ↑
             1

α<u>D</u>-Gl<u>c</u>pNAc

FIGURE 7   Structure of the dephosphorylated oligosaccharide from the serotype L3 LPS.

oligosaccharides to protein by reductive amination[23] was
not possible due to the lack of reactivity of the ketose
group of the terminal reducing 3-deoxy-$\underline{\underline{D}}$-manno-octulosonic
acid (KDO) residue and its preferential reduction by sodium
cyanoborohydride[24]. However, reductive amination of the
KDO residue could be achieved using a large molar excess of
a small molecular size spacer containing an amino group[24].
For the sake of convenience in preliminary studies this
conjugation was carried out by a previously published
procedure[25] involving the use of a 2-(4-isothiocyanato
phenyl)-ethylamine spacer. This procedure resulted in the
incorporation of from 18 to 38 oligosaccharides per
molecule of tetanus toxoid.

When injected in rabbits the conjugates produced
oligosaccharide-specific antibodies, the specificity of
which matched very closely the lipopolysaccharide-specific
antibodies used to serotype the meningococcal organisms.
An estimate of the potential of these conjugates to raise
acceptable levels of protective antibodies was obtained by
injecting them in rabbits and subjecting the rabbit
antisera to an in vitro bactericidal assay[23]. For the sake
of brevity only selected assays are shown in Table 2 but

TABLE 2   Bactericidal activities of oligosaccharide-
conjugate antisera against homologous and
heterologous meningococcal organisms.

Serotype conjugate antiserum	Bactericidal activity against strain			
	L10	L5	L2	L3
L10	1024	32	64	0
L5	8	512	0	0
L2	0	32	128	0
L3	0	0	32	4

they are representative of all the results. All the
conjugates produced rabbit antisera with bactericidal
activity predominantly against their homologous serotypes
although some minor cross bactericidal-activity was
exhibited. From the perspective of using these
oligosaccharide-conjugates as a human vaccine against group
B meningococcal disease the procedure warrants further
study. However, these preliminary results indicate that a
complex multivalent oligosaccharide-conjugate would be
required to provide complete protection against disease
caused by group B N. meningitidis.

2. CONCLUSIONS

While capsular polysaccharides are now established in
current human vaccines and will probably be used in future
vaccines, the conjugation of these polysaccharides to
protein carriers also has great potential in the future
formulation of vaccines. Of immediate benefit would be
their use in formulating a vaccine against meningitis in
infants. However, this technique could also be employed
for the general enhancement of the immune response to
polysaccharides and, to introduce into vaccines other
smaller non-immunogenic haptenic oligosaccharides found on
the surface of bacteria.

REFERENCES

1. J.B. ROBBINS, Immunochemistry, 15, 839-854 (1978).
2. H.J. JENNINGS, Adv. Carbohydr. Chem. Biochem., 41,
   155-208 (1983).
3. R. AUSTRIAN, Rev. Infect. Dis., 3(Suppl.), S1-S26
   (1981).

4. A.R. DOCHEZ and O.T. AVERY, J. Exp. Med., 26, 477-493 (1917).
5. O.T. AVERY and M. HEIDELBERGER, J. Exp. Med., 42, 367-376 (1925).
6. J.B. ROBBINS, R. AUSTRIAN, C.J. LEE, S.C. RASTOGI, G. SCHIFFMAN, J. HENRICHSON, P.H. MAKELA, C. BROOME, R.R. FACKLAM, R.H. TIESJIEMA and J.C. PARKE Jr., J. Infect. Dis., 148, 1136-1159 (1983).
7. E. KATZENELLENBOGEN and H.J. JENNINGS, Carbohydr. Res., 124, 235-245 (1983).
8. G. ZON, S.C. SZU, W. EGAN, J.D. ROBBINS and J.B. ROBBINS, Infect. Immun., 37, 89-103 (1982).
9. H.J. JENNINGS, C. LUGOWSKI and N.M. YOUNG, Biochemistry, 19, 4712-4719 (1980).
10. D.E. BRILES, C. FORMAN, S. HUDAK and J.L. CLAFLIN, J. Exp. Med., 156, 1177-1185 (1982).
11. S.C. SZU, S. CLARK and J.B. ROBBINS, Infect. Immun., 39, 993-999 (1983).
12. R. SCHNEERSON, O. BARRERRA, A. SUTTON and J.B. ROBBINS, J. Exp. Med., 152, 361-376 (1980).
13. H.J. JENNINGS and C. LUGOWSKI, J. Immunol., 127, 1011-1018 (1981).
14. E.C. BEUVERY, F. MIEDEMA, R. VAN DELFT and J. HAVERKAMP, Infect. Immun., 40, 39-45 (1983).
15. W.F. GOEBEL and O.T. AVERY, J. Exp. Med., 54, 431-436 (1931).
16. O.T. AVERY and W.F. GOEBEL, J. Exp. Med., 54, 437-447 (1931).
17. M.R. LIFELY, A.S. GILBERT and C. MORENO, Carbohydr. Res., 94, 193-203 (1981).
18. F.A. WYLE, M.S. ARTENSTEIN, D.L. BRANDT, D.L. TRAMONT, D.L. KASPER, P. ALTIERI, S.L. BERMAN and J.P. LOWENTHAL, J. Infect. Dis., 139, 52-59 (1979).
19. J. FINNE, M. LEINOREN and P.H. MAKELA, Lancet, 355-357 (1983).
20. W.D. ZOLLINGER and R.E. MANDRELL, Infect. Immun., 18, 424-433 (1977).
21. H.J. JENNINGS, A.K. BHATTACHARJEE, L. KENNE, C.P. KENNY and G. CALVER, Can. J. Biochem. 58, 128-136 (1980).
22. H.J. JENNINGS, K.G. JOHNSON and L. KENNE, Carbohydr. Res., 121, 233-241 (1983).
23. H.J. JENNINGS, C. LUGOWSKI and F.E. ASHTON, Infect. Immun., 43, 407-412 (1984).
24. R. ROY, E. KATZENELLENBOGEN and H.J. JENNINGS, Can. J. Biochem., 62, 270-275 (1984).

25. S.B. SVENSON and A.A. LINDBERG, J. Immunol. Methods,
    25, 323-335 (1979).

DISCUSSION OF HAROLD JENNING'S PAPER

M. HEIDELBERGER
Department of Pathology
New York University School of Medicine
New York, N.Y. 10016

Since this is a conference on industrial polysaccharides and because it is the antibodies to the type-specific polysaccharides that are both preventive and curative, it may not be amiss after Harold Jennings' impressive talk to make a plea for the renewed production of antipneumococcal rabbit sera as a supplement to polysaccharide vaccines.

After Goodner and Horsfall had shown good results with these antisera[1] I prepared partially purified antipneumococcal rabbit globulins for the treatment of types 1,2, and 3 pneumococcal pneumonias by half-saturation (actually 45%) with warm sodium sulfate[2]. When the globulins were adequately centrifuged it was not necessary to dialyse them as they did not contain toxic levels of sulfate. Thus sodium sulfate instead of ammonium sulfate avoided time-consuming dialysis and the risk of contamination. As soon as the infection was typed, patients received intravenously an infusion of about 70 mg of antibody nitrogen in about 100 ml of saline. Half-an-hour later a nurse brought

a few ml of the patients blood. Serum from the
clot was tested for excess antibody with a few ug
of polysaccharide and if no excess was found another
dose of antibody was given. One patient with a
massive type 3 infection required 5 bottles of antibody before an excess could be established:
recovery ensued. There was one mild urticaria, no
serum sickness, and I did not hear of any deaths.
This was the first time, I believe, that pneumonia
patients were assured of an excess of antiserum as
rapidly as possible[2].

Just then sulfadiazine and penicillin began to
supersede type-specific antiserum but there is still
a death rate from pneumococcal pneumonia. I make
this plea for some pharmaceutical company or the
pharmaceutical industry as a whole, to re-establish
rabbit colonies for the production of partially
purified type-specific antibodies that could rapidly
be sent in cases of need. Perhaps one could begin
with the pneumococcal types that are not in the
two present commercial vaccines. This might begin
as a tax-deductible philanthropy, but since people
are still dying of pneumonia there should be a steady
demand for such rapidly effective antibodies.

REFERENCES:

1. K. GOODNER, F.L. HORSFALL, JR. & F.J. DUBOS, J.
    Immunol., 33, 279-295, (1937)
2. J.C. TURNER, M. HEIDELBERGER & C.M. SOO HOO,
    Proc. Soc. Exp. Biol. & Med., 37, 734-736, (1938)

BIOMIMETIC REPROGRAPHICS

MICHAEL G. TAYLOR AND ROBERT H. MARCHESSAULT

Xerox Research Centre of Canada
2660 Speakman Drive
Mississauga, Ontario
L5K 2L1
Canada

Carbohydrate--mediated information transfer in the form of chemical recognition is the key mechanism of cell-cell interactions. This leads to adherence between bacteria (or fungi) and infected cells,a prerequisite for infection. Sugar specific adherence is best demonstrated by the carbohydrate/lectin examples. In the present study glycolipid bearing vesicles are dyed in order to be used as "toner" particles to develop a latent image which is created by imagewise denaturation of a lectin-coated surface. The adherence specificity and resolution of the carbohydrate-tagged vesicles could be demonstrated by painting images on surfaces with a lectin solution, development was achieved by simply dipping in a solution of dyed vesicles. The response time of tagged vesicles/concanavalin A interaction was followed by low angle light scattering both for the agglutination and resuspension phase of the interaction. The dynamics of the process was limited only by the mechanics of sample mixing and recording. Accordingly, a practical reprographic imaging system based on chemical recognition of biological materials appears feasible. The degree to which such a system requires an aqueous environment to enable the recognition/adherence phenomenon is still an open question.

INTRODUCTION

There is an increasing interest in the exploitation of biological systems in industrial processes. The high degree of specificity and catalytic efficiency of enzymes make them attractive as industrial catalysts. The rapidly developing science of genetic engineering promises to deliver unique organisms with properties tailored for specific functions such as the production of insulin. Although Xerox will not likely be directly involved in the development of these types of applications, the possibility of exploiting natural materials in applications of interest to Xerox should not be overlooked. This report demonstrates the use of a biological system in a novel imaging technology.

Many biological processes involve the phenomenon of chemical recognition. Chemical recognition provides a means of selecting and binding a chosen molecule with a high degree of specificity. This phenomenon is the first step of an enzyme-catalyzed reaction, where the enzyme selectively binds a particular substrate. Chemical recognition is a key feature in the functioning of the immunological system. Antibodies display great specificity in recognizing their target antigens. The application described in this report uses a lectin, a member of the family of plant proteins which are similar to antibodies in their ability to recognize and bind specific molecules. The target molecule in this case is a simple carbohydrate derivative.

This report outlines the mechanism of chemical recognition related to lectin/carbohydrate interactions. Materials considerations for use in an imaging system are then discussed. A model system for use in a reprographic application is described, with experimental data for solution and surface behaviour. Finally, some recommendations are made for further exploratory research in this area.

1.0 CHEMICAL RECOGNITION

Chemical recognition is a general term describing a variety of phenomena studied in molecular biology. In each case, a target molecule is selectively bound by another molecule (generally a protein). Examples are found in the binding of substrates by enzymes and antigens by antibodies. The key difference between chemical recognition and the types of complexation reactions generally encountered in chemistry is the high degree of specificity characteristic of the biological phenomena.

The chemical recognition phenomenon considered here is the interaction between a lectin and a carbohydrate. Lectins are members of a class of proteins found largely in plant seeds. They are defined as "naturally occuring carbohydrate-binding proteins of non-immune origin which agglutinate cells and/or precipitate complex carbohydrates" (1). Their biological activity closely mimics that of the immunological system. Lectins selectively recognize and bind particular carbohydrates but the degree of specificity varies between different lectins. The lectin used in this study, Concanavalin A, is one of the less specific lectins.

1.1 Concanavalin A (Con A)

Con A is a protein which can be isolated from Jack Beans (Canavala ensiformis). This lection was chosen for model studies since it is readily available and has been widely studied (2). In solution, the protein self-assembles into a tetramer with four identical subunits each of 27,000 molecular weight. Each subunit has one site which binds a carbohydrate. As stated above, the specificity of Con A is not absolute, however it does display the ability to recognize minor conformational differences in carbohydrates. For example, the relative binding strengths (as measured by equilibrium constant for formation of the lectin/carbohydrate complexes) are shown in Figure 1 for four closely related carbohydrates (3).

It can be seen that minor variations in molecular structure of the carbohydrates cause drastic changes in the interaction with the protein. This is in contrast to "ordinary" chemical complexation reactions which tend to be much less sensitive to these types of conformational differences. The exact nature of the complex formed when Con A binds a carbohydrate is still somewhat controversial. There is, however, a great deal of experimental data available on the interaction of Con A and other lectins with a variety of

CARBOHYDRATE	Relative Binding Strength to Concanavalin A

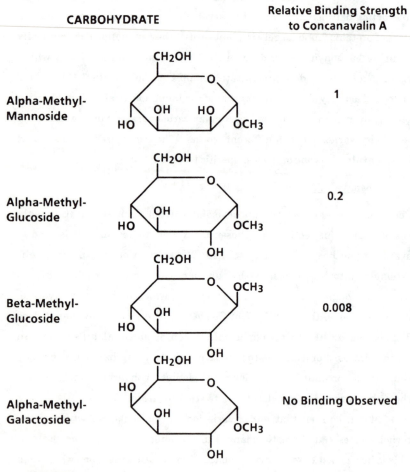

Alpha-Methyl-Mannoside	1
Alpha-Methyl-Glucoside	0.2
Beta-Methyl-Glucoside	0.008
Alpha-Methyl-Galactoside	No Binding Observed

FIGURE 1

Interaction of Concanavalin A with simple carbohydrates.

carbohydrates. Thus, systems can easily be tailored to utilize a particular carbohydrate or lectin.

1.2 Application in an Imaging System

A model imaging system is described in this report, based on the lectin/carbohydrate interaction. In this model, this interaction is used for image development, as electrostatic neutralization is used in xerography. In order to create a visible image through this interaction, one of the components in the pair must be rendered opaque. In this case, the carbohydrate component was chosen, and colored particles were prepared with surfaces labelled with a carbohydrate derivative. Phospholipid vesicles were chosen to act as the toner particles.

1.3 Vesicles

Vesicles are small (~50 nm diameter) particles formed by the dispersion of amphiphilic compounds in water. The amphiphiles are molecules with polar and non-polar molecular regions. When dispersed in water, the polar regions are readily solvated while the non-polar fragments of the amphiphile are poorly solvated. Above a critical concentration, the amphiphiles spontaneously self-assemble to form a variety of liquid crystalline mesophases. A familiar example of this behaviour is the formation of micelles by soaps.

The amphiphile used in this study was a phosphatidycholine, (lecithin) derived from hen's eggs. The structure is illustrated below. The non-polar "tail" region consists of two long chain hydrocarbon moieties. These fatty acids are typically 16-18 carbons in length. The polar "head" of the molecule is the zwitterionic phosphoryl choline function.

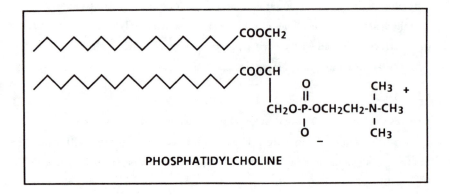

When dispersed in water, this amphiphile forms the bilayer mesophase illustrated in Figure 2. The lamellar bilayers form closed multilamellar structures often referred to as "liposomes".

When the bilayers are subjected to intense ultrasonic irradiation (sonication), the large multilamellar structures are disrupted. The liposomes are broken up into much smaller structures known as vesicles with outside diameters of approximately 50 nm. The diameter of the enclosed aqueous space in the centre of the vesicle is 20-30 nm. The sonication process is shown schematically in Figure 3.

BILAYER FORMATION

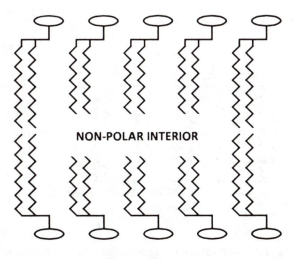

FIGURE 2

Bilayer mesophase formed by phosphatidylcholine dispersed in water.

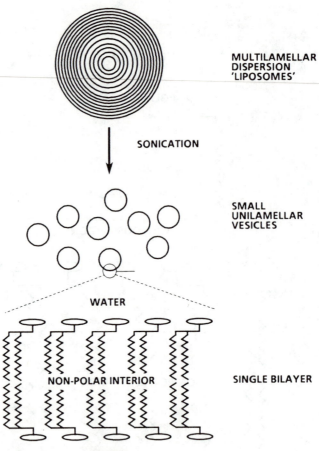

FIGURE 3

Disruption of multilamellar liposomes to form unilamellar vesicles. A magnified view of a single bilayer from a small unilamellar vesicle is shown in the bottom frame

The vesicles can be readily loaded with compatible dyes or other materials. In the present study, a hydrocarbon soluble dye, Sudan Black, was loaded into the vesicles. In order to allow recognition of the vesicles by the lectin Con A, it was necessary to label the surface of the vesicles with an appropriate carbohydrate group. This was accomplished with the glycolipid shown below.

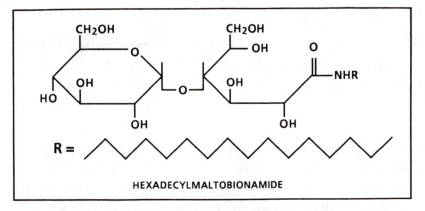

HEXADECYLMALTOBIONAMIDE

This molecule inserts into the vesicles with its long hydrocarbon chain anchored in the non polar portion of the bilayer (4,5). The carbohydrate moiety is exposed at the polar interface between the bilayer and water. Thus, the surface of the vesicle is coated with α-glucosyl moieties which can be recognized and bound by Con A. A significant amplification is achieved since the binding of the carbohydrate moiety by Con A immobilizes the entire vesicle and its contents. The final component of the vesicles is shown below.

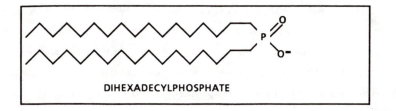

DIHEXADECYLPHOSPHATE

This component was added to impart a negative surface charge to the vesicles. It was previously reported (4) that the vesicle/Con A interaction requires a negative surface charge on the vesicles. In addition, some stabilization of the vesicle suspension is achieved, since particle agglomeration is hindered by electrostatic repulsion between vesicles.

2.0 SOLUTION STUDIES

The behaviour of these components was first examined in solution. The dyed and carbohydrate-tagged vesicles form a highly colored, clear suspension. Addition of a small amount of Con A solution to the vesicle suspension causes a rapid and dramatic change in the suspension. A highly colored precipitate forms, leaving a colorless supernatant. Addition of a simple sugar (*e.g.* maltose) to the precipitate, followed by mixing, redisperses the vesicles yielding a clear, colored suspension.

The mechanism explaining this behaviour is shown schematically in Figures 5 and 6. The initial precipitation is shown in Figure 5. Each Con A assembly consists of a tetramer with four carbohydrate binding sites. The vesicle surfaces have a high concentration of the carbohydrate taggant which is recognized and bound by Con A. A large network of Con A/vesicle complexes can form because of the multi-valent characteristics of each component. Each Con A can bind one or more vesicles which can then be bound by a different Con A assembly leading quickly to the formation of a large network which is insoluble.This phenomenon is referred to as agglutination.

The network formation is readily reversed by addition of a solution containing an excess of a low molecular weight carbohydrate which is bound by Con A. As shown schematically in Figure 6, the free carbohydrate displaces the carbohydrate-tagged vesicles from the binding sites of the Con A. The vesicles are thus re-suspended, and the Con A/carbohydrate complexes formed are soluble.

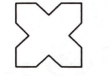

CONCANAVALIN A

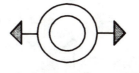

**CARBOHYDRATE-TAGGED
VESICLES**

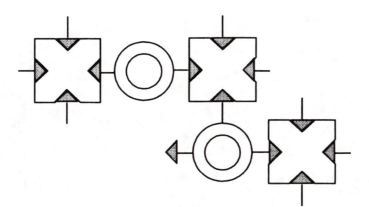

**AGGLUTINATION
THREE DIMENSIONAL NETWORK FORMS
AND PRECIPITATES FROM SOLUTION**

FIGURE 5

Interaction between Concanavalin A and carbohydrate-tagged
vesicles

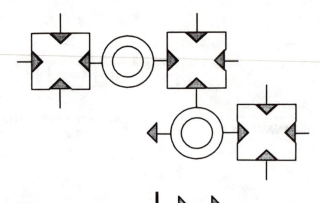

**ADDITION OF FREE SUGAR
TO AGGLUTINATED VESICLES**

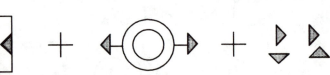

**SOLUBLE SUGAR-
CONCANAVALIN A
COMPLEX**
    FREE VESICLES
    EXCESS FREE SUGAR

FIGURE 6

Reversal of vesicle agglutination by competitive binding of free sugars.

2.1 Light Scattering

The agglutination-resuspension can be monitored by following the light scattering properties of the solution. Optical studies of vesicle agglutination generally employ measurements of light transmission through the suspension. As the agglutinating complexes form, the intensity of light transmitted through the supsension decreases. This method was not feasible in the present case due to the presence of the Sudan Black dye in the vesicles. This dye absorbs intensely over a wide portion of the spectrum. Preliminary observations showed that the influence of agglutination on the light transmission was much smaller than that due simply to absorbance by the dye. Measurements of scattered intensity did allow monitoring of the agglutination process in the presence of the dye.

The increase in particle size resulting from the agglutination of the tagged vesicles and Con A could be monitored through the increase in forward light scattering of the vesicle suspension. The increase in scattered intensity with particle growth is greatest for scattering angles close to the incident beam. The forward scattering ($\lambda = 440$ nm) at $10°$ from incident (I_{10}) was followed in suspensions of tagged vesicles to which Con A had been added.

Figure 7 illustrates the growth in particle size after addition of Con A to a suspension of tagged vesicles. A very dilute suspension of vesicles was employed so that the agglutination proceeded at a rate which could be readily followed. In addition, the suspension was stirred between measurements to avoid settling of the agglutinated particles. After ~45 minutes, the agglutination had approached equilibrium. At this point, a solution containing excess maltose was added. This carbohydrate competes with the α-glucosyl moiety of the glycolipid taggant of the vesicles for Con A binding sites. As the vesicles are displaced from the Con A, the agglutinated network breaks up. The decrease in particle size is shown by the drastic decrease in forward light scattering.

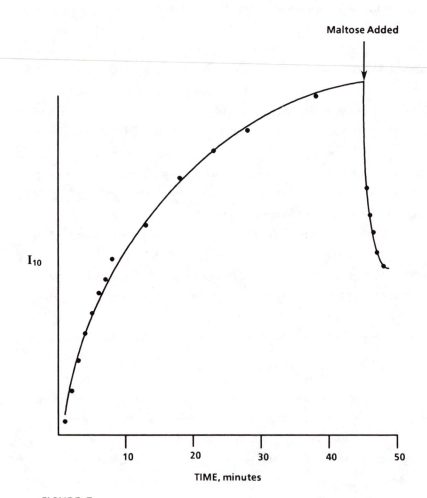

FIGURE 7

Variation in scattered light intensity after addition of Con A to vesicle suspension. Maltose was added 45 minutes after addition of Con A.

This experiment provides a ready means for monitoring the agglutination process. The use of forward light scattering was necessary in this case due to the strong dye absorbance of the vesicles. This technique should be applicable for vesicles containing any absorbing components, or for other types of particles used in agglutinating systems *(vide infra)*.

3.0 IMAGE DEVELOPMENT

In the xerographic process, image development is achieved via electrostatic neutralization of a charged site on a photoreceptor by oppositely charged toner particles. In the present model, chemical recognition is substituted for electrostatic neutralization as the image-producing mechanism. In a recognition based imaging system, it is necessary to fix one of the components to a substrate where it can be developed by interaction with the second component. In the present study, the lectin was bound to a surface.

There are several possible approaches to the problem of preparing a two dimensional substrate coated with lectin. One approach involves covalent attachment of the lectin to the surface. The chemistry of covalently binding proteins, such as lectins, to solid surfaces is known for many applications in the field of immobilized enzyme technology. In the present study, however, surface coating was achieved by preparing a gelatin based emulsion containing the lectin on a smooth substrate.

An emulsion was chosen as the means of surface coating since it provides a hydrated medium for the lectin. The chemical recognition phenomenon requires a conformational match between the protein lectin and the carbohydrate. Since the protein conformation is likely influenced by the presence of water, a method of immobilization which maintained an aqueous medium for the lectin was preferred.

The image development process is outlined schematically in Figure 8. The lectin is bound to the surface and presents its carbohydrate-binding sites to the external medium. Exposure of the surface to a suspension of tagged vesicles leads to binding of the vesicles at surface sites occupied by the lectin. Since the tagged vesicles also contain a dye, the site is visibly marked.

The experimental result is shown in Figure 9 . A cellulose acetate strip was coated with an emulsion composed of Con A dispersed in gelatin. The emulsion was then allowed to dry and fix on the surface. When the acetate strip was dipped into a vesicle suspension, the vesicles uniformly marked the acetate surface bearing the emulsion. A rudimentary image was produced by "painting" an "X" on a cellulose acetate surface with gelatin/Con A emulsion. This formed a latent image which was subsequently developed by dipping the acetate sheet into a suspension of dyed, tagged vesicles. As can be seen in Figure 9, image development occurred via chemical recognition at the sites where the latent image was produced.

4.0 FURTHER RESEARCH DIRECTIONS

4.1 Latent Image Formation

The ability to develop is only one component of an imaging process. A means of producing the latent image is also required. In this study, a latent image was produced simply by mechanically depositing emulsion on a substrate in the shape of the desired image. In principle, this method could be applied to produce latent images. Such a system would employ jetting as in ink jet printing to apply the emulsion to a plain surface substrate in the form of a latent image. Development would occur by exposing the surface to a suspension of tagged toner particles.

Other methods of producing latent images are possible which exploit the characteristics of chemical recognition. The phenomenon depends strongly on a conformational match between the two components. The recognition ability of the protein component can readily be destroyed by

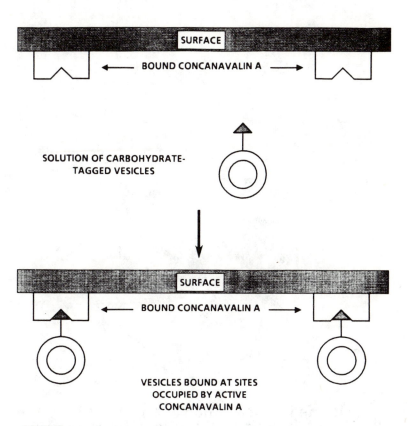

FIGURE 8

Schematic representation of binding of tagged vesicles to substrate surface. Binding occurs at sites occupied by active lectin.

Figure 9
Image formed by deposition of Con A/gelatin with subsequent development by tagged, dyed vesicles.

denaturation, *i.e.* by altering its native conformation. Proteins can be denatured by, for example, exposure to heat, light or electron beams. A possible application involves coating the surface of a substrate uniformly with lectin. The lectin on the surface could then be denatured image-wise using for example a modulated laser. At locations heated by the laser radiation, the recognition activity of the lectin would be destroyed. Development of the substrate by exposure to tagged particles would result in marking only at sites untouched by the laser. The resolution in such a system could be quite high since the proteins would be denatured at the molecular level. Thus, resolution would be limited by the beam width. The use of an electron beam for denaturation could lead to very high resolution imaging.

It is also possible to envisage an imaging system with the locations of the two recognition components reversed. In this case, the toner particles would be tagged with the protein component. In this analog of the present model, the substrate would be coated with the appropriate carbohydrate. Latent image formation would then require modification of the carbohydrate on the substrate, followed by development by exposure to the lectin-tagged particles.

4.2 Other Materials

This model system employed the lectin/carbohydrate pair as the two imaging components. Similar applications are possible using antibody/antigen pairs. The unique feature of this system is the possibility of preparing custom antibodies. The technology available in the field of monoclonal antibodies allows production of specific antibodies which are active against a selected antigenic molecular species. It is also possible to conceive of a system based on the recognition of a substrate by an enzyme.

Vesicles were used as "toner" particles in the present study. Vesicles are attractive in this application since they can be readily surface-tagged with an appropriate label. In addition, a variety of materials can be carried by the vesicles. The present application employed an organic dye solubilized in the hydrocarbon region of the vesicles. Vesicles were also prepared using an organic-soluble Ferrofluid. These vesicles showed poor physical stability on standing, but were interesting due to their magnetic properties. It is also possible to load materials into the internal aqueous space of the vesicles. Vesicles loaded in this fashion with drugs are presently being studied. Surface labelled vesicles can be directed to specific sites in an organism in a manner similar to the application discussed in this report (6).

5.0 SUMMARY AND CONCLUSIONS

The application of a rudimentary imaging system based on chemical recognition has been demonstrated. In its present form, this system is very crude, but warrants further examination. The potential benefits of a recognition-based imaging system can be summarized:

5.1 Specificity

Chemical recognition provides a much higher degree of specificity than is observed in other chemical and physical phenomena. It may be possible to exploit this specificity in specialised aplications such as color, for example by using a system similar to that described above with separate lectin/carbohydrate pairs for each color.

5.2 Resolution

The use of a molecular interaction in a imaging system should allow very high resolution. The achievable resolution will be limited by the method used to form the latent image. Image-wise denaturation by electron beam is an example of a potential method for very high resolution marking.

5.3 Versatility

The present model system dye-loaded vesicles as smart particles. This system could be readily modified by loading vesicles with other materials in either the aqueous inner space or in the non-polar region as in the present study. Other materials such as latex beads or lectin-coated colloidal metal particles (7) could be used in place of the vesicles.

As mentioned above, there are several unresolved questions regarding the feasiblity of an imaging system based on chemical recognition. The resolution of these questions will require further research in both the fundamental and applied aspects of this phenomenon. We intend to address the following questions which remain open:

5.4 Hydration

In the present study, both components of the recognition pair were used in hydrated media *i.e.* solution or gel. It would be useful to know whether a similar system could be achieved using components in the dry state. Room humidity may be sufficient to permit the conformational matching required for the recognition interaction.

5.5 Surface Binding Kinetics

Development of a latent image by recognition may be the rate-limiting factor in process speed. Development speed for a liquid based system is not expected to be lower than that of other liquid development processes, *e.g.* Versatec electrostatic imaging.

6.0 ACKNOWLEDEMENTS

We are grateful to Dr. Michael Hair and Professor John Baldeschweiler for many useful discussions.

7.0 MATERIALS AND METHODS

Concanavalin A, egg phosphatidylcholine, dihexadecyl phosphate and Sephadex G-50-80 were purchased from Sigma. Gelatin, (Laboratory Grade 275 Bloom), and Rexyn 101 ion exchange resin were purchased from Fisher.

Technical grade hexadecylamine was purchased from Kodak. Tris (hydroxymethyl) methyl ammonium chloride was purchased from BDH Chemicals. The other reagents employed were of analytical reagent grade.

The synthetic glycolipid, hexadecylmaltobionamide, was prepared according to a published procedure (8) with some modifications. Difficulties were encountered in the preparation of potassium maltobionate following the procedure of Moore and Link (9). The product formed an oil or tar and I_2 was entrained in the solid. This procedure was modified to avoid the presence of water in the reaction mixture.

Thus, 8 g maltose was dissolved in 100 ml CH_3OH with heating and the solution was added to a solution of 10 g I_2 in 150 ml CH_3OH. Dropwise addition of methanolic KOH (10 g KOH/100 ml CH_3OH) was begun immediately. A precipitate formed in the stirred solution. After the addition of 100 ml methanolic KOH, the addition was stopped and the suspension was stirred for ~10 minutes. The remaining 100 ml KOH solution was then added dropwise. The suspension was lightly coloured after complete addition of the KOH. The stirring was continued until the suspension was essentially colourless (~30 minutes). The stirring was then continued for ~15 minutes in an ice bath. The suspension was then filtered under suction and the solids were washed with cold CH_3OH and then air-dried. The crystals were dissolved in H_2O and added to a slurry of 100 g Rexyn 101 ion exchange resin which had been washed with 1N HCl and H_2O. The final aqueous volume was 100 ml. The slurry was stirred for 30 minutes to convert potassium maltobionate to free maltobionic acid.

The aqueous solution was filtered from the slurry and evaporated to near-dryness under reduced pressure at 50°C. Methanol (100 ml) was added and the solution was evaporated to dryness yielding a syrup. The maltobiono-1,5-lactone was formed by dissolving the syrup in 100 ml CH_3OH and evaporating the solvent on a rotary evaporator at 50°C. The dissolution/evaporation cycle was repeated 5 more times. The lactone forms a solid foam as the solvent is removed. The yield was 7 g. The lactone was then dissolved in 25 ml CH_3OH and was added to a warmed solution of 5 g hexadecylamine in 25 ml CH_3OH. The solution was then stirred and a precipitate formed after ~30 minutes. The stirring was continued overnight, and the precipitate was separated by centrifugation. The solids were crystallized three times from absolute ethanol. The crystals were then filtered, washed with ~20 ml CH_3OH and air-dried. The crystals were dissolved in 100 ml H_2O with heating and the solution was then frozen at dry ice temperature and freeze-dried overnight. The yield of freeze-dried material was 3.8 g. The synthetic procedure is outlined in Figure 10.

7.1 Preparation of Vesicles

The vesicle components; egg phosphatidylcholine (100 mg), hexadecylmaltobionamide (78 mg), dihexadecylphosphate (71 mg) and Sudan Black B (100 mg) were mixed in 100 ml CH_2Cl_2. The solvent was evaporated to dryness under reduced pressure and the dry material was kept under water aspirator vacuum for a further 30 minutes. The lipids were suspended in 20 ml buffer (50 mM Tris, 0.5 mM Ca^{2+}, 0.5 mM Mn^{2+}, pH 7.1) with vigorous agitation on a Vortex mixer. Glass beads (3 mm diameter) were added to the flask to aid in removing the lipid film from the flask surface. The suspension was then sonicated to produce vesicles (Heat Systems Sonicator, output setting 6, 50% duty cycle, 5 minutes). The suspension was then centrifuged in a clinical centrifuge and the supernatant was separated from a pellet of solid dye which had not been solubilized. The suspension was then passed through a Sephadex column (1 g Sephadex G-50-80 swollen in buffer). The resulting vesicle suspenson was clear and highly coloured.

Magnetic vesicles were also prepared. The vesicles were formed as described above, with the exception that the Sudan Black was replaced by 1 ml hydrocarbon based Ferrofluid (Ferrofluidics Corp.) and the aqueous volume was increased to 100 ml.

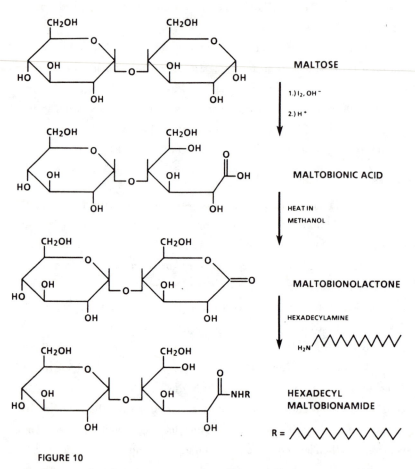

FIGURE 10

Synthetic route for the preparation of Hexadecylmaltobionamide

7.2 Preparation of Gelatin emulsion

A 6% solution of gelatin in water prepared by heating. The solution was allowed to cool and 1 ml of the solution was then added to 1 ml Concanavalin A solution (1 mg Con A/ml buffer). The resulting solution was immediately cast on a cellulose acetate sheet and was spread over the surface. Air drying at room temperature yielded a transparent emulsion.

8.0 REFERENCES

1.   I. J. GOLDSTEIN, private communication.
2.   Concanavalin A as a Tool, A. I. Bittiger, H. H. Schnebli, eds., John Wiley & Sons Ltd. N.Y., 1976.
3.   I. J. Goldstein, Chapter 4 in reference 2.
4.   R. Y. HAMPTON, R. W. HOLZ and I. J. GOLDSTEIN, J. Biol. Chem., 255, 6766-6771, (1980).
5.   H. BADER, H. RINGSDORF and J. SKURA, Angew. Chem. Int. Ed. Engl., 20, 91-92, (1981).
6.   M. R. MAUK, R. C. GAMBLE, and J. D. BALDESCHWIELER, Science, 207, 209-311, (1980).
7.   J. ROTH, J. Histochem. Cytochem., 31, 987-999, (1983).
8.   T. J. WILLIAMS, N. R. PLESSAS and I. J. GOLDSTEIN, Arch. Biochem. Biophys., 195, 145-151, (1979).
9.   S. MOORE and K. P. LINK, J. Biol. Chem., 133, 293-311, (1940).

CONSIDERATION OF POLYSACCHARIDES FOR INDUSTRIAL USE

ROY L. WHISTLER
Department of Biochemistry
Purdue University
West Lafayette, IN 47907

In much of the work with polysaccharides investigators have needed to react the polymers with chemicals or enzymes to make derivatives or to demonstrate structures. In so doing it is important to be aware of the polysaccharide's chemical reactivity, its physical nature and its association with other molecules. In this brief overview I discuss several polysaccharides that are commonly met in industrial practice and point out considerations that must be kept in mind in any direct use of these polymers if their structures are to be maintained and especially if their molecular weights are to be maintained at high levels.

Investigators often react polysaccharides chemically with the belief that uniformity of derivatization is attained and is attained without depolymerization. It is most important to recognize that uniform derivatization is very difficult to attain and, perhaps, still more difficult is the derivatization of large polysaccharides without some, and often extensive chain cleavage. Too often an investigator reacts a polysaccharide under derivatizing

conditions that are unrecognized as harsh and concludes
erroneously that if the derivatization is incomplete it
is uniform and that no chain length cleavage has occurred.
These difficulties arise from occurrence of polysaccharides
in close molecular association, either attached to each
other by strong intermolecular bonding due, usually, to
enormous numbers of hydrogen bonds, or by the polysaccharides intermolecular attachment through secondary bonds or
even covalent bonds to other molecules in the natural biological material. It must be born in mind also that the
often rigorous treatment necessary to effect derivatization causes some breakage of glycosidic bonds resulting
in molecular depolymerization. In dealing with macromolecules it must be remembered that only a minor cleavage
of a chain greatly effects molecular behavior. In viscosity measurement alone an enormous change is evident.
Thus, but one chain cleavage in a large polysaccharide,
if the cleavage is near the center, will lower the molecular weight by fifty percent but can decrease the viscosity to 12 percent of its initial value.

1. Starch

First let us look briefly at starch. It is unique among
all other polysaccharides because it occurs in the form
of discrete packages, the starch granules.

        With initiation of starch biosynthesis α-$\underline{D}$-gluco-
pyranosyl units begin to be joined uniformly by 1-->4-
linkages. These chains grow and are continually length-
ened. However, most of the chains, when they reach 35-50
units in length, are cleaved by a branching enzyme, a
transferase, and joined to another chain by a 1-->6-

linkage to produce the initial branching in amylopectin or
to increase the branching in an already initiated amylopectin molecule. In any case the synthase enzyme continues
to increase the length of the linear amylose molecules, or
the length of the branches in the outer limbs of amylopectin molecules. Some investigators believe that amylose
molecules are slightly branched, having 2 to 5 branches per
1000 α-$\underline{D}$-glucopyranosyl units. If branches do occur they
must be few and quite long because amylose by itself or in
the form of derivatives, such as the peracetate, can form
films and fibers of a strength equal to, and of a flexibility greater than, comparable cellulose. The conceived
nature of amylopectin branching has undergone change over
the years. Originally the bush-shaped or tree-shaped structure of K.H. Meyer was accepted, and remains correct in the
judgement of many modern investigators. However, Professor
Zero Nikuni has suggested that a more exact structure is
one which I call a tassels-on-a-string structure, but
called a chain of clusters by Dexter French. Here short
chains of 12-15 $\underline{D}$-glucopyranosyl units occur about every
25 units of the main chain of over a thousand units. Continued work on the structure of amylopectin shows that
amylopectin on enzymatic hydrolysis of its 1-->6 linkages
produces fractions of degrees of polymerization of 45-50,
18-20, and 10-12. No oligosaccharides lower than malto-
hexose are found, indicative of the shortest branches
present.

     In the early biosynthesis of starch molecules there
soon occurs a spontaneous nucleation and insolubilization
at a point called the hilum. From this point continuous
formation and precipitation of starch molecules occurs in

a radical fashion with production of a spherocrystalline-like structure. Amylose molecules are apparently mixed among amylopectin molecules. Although some amylose molecules may remain individual, most seem to be associated as double helices in anti-parallel arrangement and often the double helices seem to be associated into small groups.

Most granules grow rapidly from the hilum, although it is not centrally located in most granules. Granular growth is not usually continuous but, occurs in surges due to the diurnal variation in sugar supply from the photosynthetic process. As a result most granules show growth rings or lamella. Professor Dexter French suggested that an amylopectin molecule could have its beginning, or reducing end, at the inner edge of a lamella and continue outward to the peripheral edge of the lamella. At the termination of development, the granule has no detectable encompassing membrane but only the closely packed non-reducing chain ends from amylose molecules and the abundant non-reducing ends of amylopectin branches. The only protein on the surface of the granules appears to be fragmentary or lacy remains of the synthesizing enzymes, some of which are still rather tightly bound at some locations on the granular surface.

As a consequence of the radial spherocrystalline growth, a granule in a polarizing microscope shows a polarization diffraction cross of isocline, or a maltese cross, with the center at the hilum. Crystallinity demonstrated in x-ray patterns show cereal starches evidencing an A-pattern that indicates chains of amylose in antiparallel double helices separated by insterstitial water. Tuber and root starches produce B-patterns with

water in sheets and columns up to 30% of that present.
These patterns are formed from uncomplexed amylose molecules or long lengths of otherwise structured chains. Amylopectin molecules seem to be positioned with their reducing ends inward as would be expected from an outwards biosynthetic growth of chains and branches.

Crystallinity is also evident from measurements in a
scanning calorimeter which reflects the break-up of crystalline regularity on heating and depicts the melting
energy required.

Undamaged starch granules are not soluble in cold
water, but can reversibly imbibe water and swell slightly.
The percent increase in granule diameter ranges from 9.1%
for normal corn starch to 22.7% for waxy starch. However
as the temperature is increased, the starch molecules
vibrate so extensively that they break intermolecular
bonds and allow their hydrogen bonding sites to engage
more water molecules. This penetration of water, and the
increased separation of more and more segments of starch
chains, increases randomness in the general structure and
decreases the number and size of crystalline regions.
Continued heating in the presence of abundant water results in a complete loss of crystallinity as judged by
loss of birefringence the nature of the x-ray pattern.
The point at which birefringence first disappears is regarded as the gelatinization point, or gelatinization
temperature. It usually occurs over a narrow temperature
range with larger granules gelatinizing first and smaller
granules later, although this is not a universal pattern.
Other methods for measuring gelatinization involve measuring loss of turbidity, increase solubility, dye absorp-

tion, enzyme action, chemical reactivity, or changes in x-
ray pattern or nuclear magnetic resonance. Of these, one
of the most sensitive and one easy to measure is the increase in extent of enzyme hydrolysis using glucoamylase,
or a mixture of alpha amylase and glucoamylase wherein
produced D-glucose is determined by glucose oxidase.

In normal commercial processing starch granules
quickly enlarge past the reversible point. Water molecules enter between starch chains, break interchain bonds
and establish hydration layers around starch molecules to
effect a lubrication of chains which become more and more
fully separated and individually solvated. At first the
entrance of water causes a tangential swelling pressure
that induces granule enlargement so that the granules increase hundreds of times in size, forcing some peripheral
molecules to flow together to reestablish intermolecular
bonding to produce a rather delicate outer membrane as
a artifact. In a gently stirred and heated 5% starch
suspension granules will imbibe water until filling the
container to produce a highly viscose gel, a starch paste.
The highly swollen granules can be easily broken and dis-
integraded by stirring to cause a large decrease in viscosity.

In swelling of starch granules the hydrated linear
amylose molecules can more easily diffuse through the membrane to accumulate preferentially in the external water
phase. This partial fractionation of starch molecules is
responsible for some aspects of paste behavior.

These various effects in starch swelling are seen in
a Brabender amylograph where viscosity is continuously
recorded as temperature is constantly raised. At the

peak, or gelatinization point some of the granules have been broken due to stirring and if stirring is continued the viscosity decreases. On cooling starch molecules, reassociate to form a gel the firmness of which depends upon how much interference occurs from other ingredients that may be present.

With this background it is easy to understand that derivatization or other chemical reactions applied to starch in the granular form will not produce uniform nor complete reaction. In the commercial derivatization of starch or in commercial oxidation, starch is reacted in granular form with low levels of reagent and with reaction or derivatization taking place irregularly and with most of it occurring in the amosphorous regions. While such reaction is quite satisfactory for most commercial products it is not satisfactory for most research purposes or to produce a uniform product that still retains most of the molecular chains of the same lengths as found in the natural granule. For uniformity of reaction without chain cleavage special solvents are needed to disperse the starch molecules and to conduct the reaction under mild nonforcing conditions.

2. Cellulose and Other Polysaccharides

Cellulose molecules are even more tightly bound intermolecular than are starch molecules. Hence, for their uniform reaction they must be dispersed carefully in nondegradative solvents and reacted in solvents or in amorphorous masses where each of the $\underline{D}$-glucopyranosyl units is equally available.

Less care needs to be used for gums, hemicelluloses and other polysaccharides containing side chains or

branches that inhibit intermolecular association of the
macromolecules.   But even in these instances molecular
association of the macromolecules.   But even in these instances molecular splitting easily occurs and it is difficult to obtain uniform reaction without depolymerization.

Many derivatives given in the literature are suspect
both as to uniformity of product and to molecular weight
values approaching the natural polymer.

PROGRAM

Chairman: R. L. Whistler

Session A: May 31, 1984; 8:30 AM to 12:10 PM

Opening Address: V. Crescenzi

Early History of Polysaccharide Chemistry: *H. Morawetz*

NEW CARBOHYDRATE POLYMERS

Capsular Polysaccharides of Gram-Negative Bacteria:
G. G. S. Dutton

Structural and Rheological Studies of the Extracellular Polysaccharides from Bacillus Polymyxa:
I. C. M. Dea

Oligosaccharide Fragments of Polysaccharides with
Commercial Potential: *P. Albersheim*

Round Table Discussion: Leader, *G. O. Aspinall*

Session B: May 31, 1984; 1:30 PM to 4:30 PM

BULK PROPERTIES AND INDUSTRIAL
UTILIZATION OF POLYSACCHARIDES

Celluloses and Amyloses: A Review of Structures and
Properties: *A. Sarko*

Molecular Interactions in Polysaccharides and their
Relationship to Bulk Properties: *W. T. Winter*

Polysaccharide Processing Using Membrane Separations:
J. R. Vercellotti

Round Table Discussion: Leader, *D. R. Durso*

Session C: June 1, 1984; 8:30 AM to 12:00 Noon

INCREASING AND DIVERSIFYING
PERFORMANCES OF POLYSACCHARIDES
BY CHEMICAL AND ENZYMATIC
MODIFICATIONS

Selective Chemical Functionalization of Polysaccharides: *D. Horton*

New Derivatives of Chitin and Chitosam: Properties and Applications: *A. Muzzarelli*

Preparation and Characterization of Enzymatically Derived Oligosaccharides and Segments from Glycosaminoglycans: *M. K. Cowman*

New Drugs from Heparin: *B. Casu*

Round Table Discussion: Leader, *M. W. Rutenberg*

Session D: June 1, 1984; 1:30 to 4:15 PM

NEW PHENOMENA AND OPPORTUNITIES
BASED ON CHEMICAL RECOGNITION
INVOLVING CARBOHYDRATES

Polysaccharides and Conjugated Polysaccharides as Human Vaccines: *H. J. Jennings*

Carbohydrate Mediated Recognition, Targeting and Marking: *R. H. Marchessault*

Round Table Discussion: Leader, *R. L. Whistler*

Closing Remarks: S. S. Stivala

LIST OF WORKSHOP PARTICIPANTS

Name	Affiliation	Country
P. Albersheim	Department of Chemistry, University of Colorado, Boulder,	U.S.A.
G. O. Aspinall	Department of Chemistry, York University, Downsville, Ontario,	Canada
D. A. Baker	Chembiomed. Ltd., University of Alberta, Edmonton,	Canada
L. Benzing Purdie	C.B.R.I. Agriculture, Ottawa, Ontario,	Canada
R. W. Binkley	Cleveland State University, Cleveland, Ohio,	U.S.A.
A. Bradbury	General Foods Corporation, White Plains, New York,	U.S.A.
G. L. Brode	Union Carbide Corporation, Bound Brook, New Jersey,	U.S.A.
J. Carpenter	Hershey Foods Corporation, Hershey, Pennsylvania,	U.S.A.
B. Casu	Institute "G. Ronzoni" Milan	Italy
J. Catcher	General Foods Corporation, Cranbury, New Jersey,	U.S.A.
A. Cesaro	Institute of Chemistry, University of Trieste, Trieste,	Italy
C. K. Chiklis	Polaroid Corporation, Cambridge, Massachusetts,	U.S.A.
M. K. Cowman	Department of Chemistry, Polytechnic Institute of New York, Brooklyn,	U.S.A.
V. Crescenzi[*]	Department of Chemistry, University "La Sapienza", Rome,	Italy
I. C. M. Dea[**]	Unilever Research, Colworth Laboratory, Bedford,	U.K.
D. F. Durso	Johnson and Johnson Co., East Windsor, New Jersey,	U.S.A.
G. G. S. Dutton	Department of Chemistry, British Columbia University, Vancouver,	Canada

Name	Affiliation	Country
M. Fishman	U.S. Department of Agriculture, Philadelphia,	U.S.A.
R. B. Friedman	American Maize Products Co., Hammond, Indiana,	U.S.A.
C. Fulger	General Foods Corporation, White Plains, New York,	U.S.A.
I. Furda	General Mills Inc., JFB Technical Center, Minneapolis,	U.S.A.
M. Heidelberger	Department of Pathology, New York University Medical Center, New York,	U.S.A.
D. Horton	Department of Chemistry, Ohio State University, Columbus,	U.S.A.
C. T. Hou	Exxon Research, Annandale, New Jersey,	U.S.A.
H. J. Jennings	National Research Council, Ottawa, Ontario,	Canada
F. T. Jones	Department of Chemistry and Chemical Engineering, Stevens Institute of Technology, Hoboken, New Jersey,	U.S.A.
E. Just	Hercules Inc., Wilmington, Delaware,	U.S.A.
A. H. King	Elinor H. King, Westfield, New Jersey,	U.S.A.
S. Levine	T.J. Lipton Inc., Englewood Cliffs, New Jersey	U.S.A.
T. Lindstrom	General Foods Corporation, Tarrytown, New York,	U.S.A.
R. H. Marchessault	Xerox Research Center, Mississauga, Ontario,	Canada
P. P. Mazzella	Department of Chemistry, College of Staten Island, New York,	U.S.A.
D. G. Medcalf	Hershey Foods Corporation, Hershey, Pennsylvania,	U.S.A.
M. S. Miller	Kraft Inc., Glenview, Illinios	U.S.A.
R. Moorhouse	Kelco Division of Merck, San Diego, California,	U.S.A.
H. Morawetz	Department of Chemistry, Polytechnic Institute of New York, Brooklyn,	U.S.A.

Name	Affiliation	Country
A. Muzzarelli	Department of Biochemistry, University of Ancona,	Italy
E. Nestaas	Petroferm Research, Massachusetts,	U.S.A.
S. Paoletti	Institute of Chemistry, University of Trieste,	Italy
M. Petitou	Institute Choay, Montrouge,	France
P. Pfeffer	U.S. Department of Agriculture, Philadelphia,	U.S.A.
L. Z. Pollara	Polymer Processing Institute, Stevens Institute of Technology, Hoboken, New Jersey,	U.S.A.
J. Reuben	Hercules Inc., Wilmington, Delaware,	U.S.A.
E. Robinson	Genzyme Corporation, Boston, Massachusetts,	U.S.A.
M. W. Rutenberg	National Starch and Chemical Corporation, Bridgewater, New Jersey,	U.S.A.
A. Sarko	Department of Forestry, State University of New York, Syracuse, New York,	U.S.A.
P. Sinay	Laboratoire Biochemie Structurale, Orleans,	France
D. Slobodin	Lonza Inc., Fairlawn, New Jersey,	U.S.A.
C. A. Snyder	General Foods Corporation, White Plains, New York,	U.S.A.
H. Stahl	General Foods Corporation, White Plains, New York,	U.S.A.
K. D. Stanley	A.E. Staley Mfg. Co., Decatur, Illinios,	U.S.A.
N. F. Stanley	Marine Colloids Division of FMC, Rockland, Maine,	U.S.A.
A. J. Stepanovic	Texaco Research Center, Beacon, New York,	U.S.A.
S. S. Stivala	Department of Chemistry and Chemical Engineering, Stevens Institute of Technology, Hoboken, New Jersey,	U.S.A.
H. Termeer	Genzyme Corporation, Boston, Massachusetts,	U.S.A.

WORKSHOP PARTICIPANTS

Name	Affiliation	Country
J. Tsai	National Starch and Chemical Corporation, Bridgewater, New Jersey,	U.S.A.
J. R. Vercellotti	V Labs Inc., Covington, Louisiana,	U.S.A.
C. C. Wan	Texaco Research Center, Beacon, New York,	U.S.A.
R. L. Whistler	Department of Biochemistry, Purdue University, West Lafayette, Indiana,	U.S.A.
G. Whitesides	Genzyme Corporation, Boston, Massachusetts,	U.S.A.
W. T. Winter	Department of Chemistry, Polytechnic Institute of New York, Brooklyn,	U.S.A.
M. Yalpani	B.C. Research, Vancouver, British Columbia,	Canada
H. F. Zobel	CPC International, Argo, Illinios,	U.S.A.

[*] Visiting Professor of Chemistry, Stevens Institute of Technology, Academic Year, 1983-1984.

[**] Visiting Professor of Chemistry, Polytechnic Institute Brooklyn, New York, Academic Year 1983-1984.